Coding Theory and Number Theory

Mathematics and Its Applications

Volume 554

Coding Theory and Number Theory

by

Toyokazu Hiramatsu

Hosei University,
Tokyo, Japan

and

Günter Köhler

Würzburg University,
Würzburg, Germany

KLUWER ACADEMIC PUBLISHERS
DORDRECHT / BOSTON / LONDON

A C.I.P. Catalogue record for this book is available from the Library of Congress.

ISBN 978-90-481-6257-4

Published by Kluwer Academic Publishers,
P.O. Box 17, 3300 AA Dordrecht, The Netherlands.

Sold and distributed in North, Central and South America
by Kluwer Academic Publishers,
101 Philip Drive, Norwell, MA 02061, U.S.A.

In all other countries, sold and distributed
by Kluwer Academic Publishers,
P.O. Box 322, 3300 AH Dordrecht, The Netherlands.

Printed on acid-free paper

Contents

Preface

This book grew out of our lectures given in the Oberseminar on 'Coding Theory and Number Theory' at the Mathematics Institute of the Würzburg University in the Summer Semester, 2001. The coding theory combines mathematical elegance and some engineering problems to an unusual degree. The major advantage of studying coding theory is the beauty of this particular combination of mathematics and engineering. In this book we wish to introduce some practical problems to the mathematician and to address these as an essential part of the development of modern number theory.

The book consists of five chapters and an appendix. Chapter 1 may mostly be dropped from an introductory course of linear codes. In Chapter 2 we discuss some relations between the number of solutions of a diagonal equation over finite fields and the weight distribution of cyclic codes. Chapter 3 begins by reviewing some basic facts from elliptic curves over finite fields and modular forms, and shows that the weight distribution of the Melas codes is represented by means of the trace of the Hecke operators acting on the space of cusp forms. Chapter 4 is a systematic study of the algebraic-geometric codes. For a long time, the study of algebraic curves over finite fields was the province of pure mathematicians. In the period 1977 – 1982, V. D. Goppa discovered an amazing connection between the theory of algebraic curves over finite fields and the theory of q-ary codes. This created a much stronger interest in the area. The idea is quite simple and generalizes the well known construction of RS-codes. The latter uses polynomials in one variable over finite fields, and Goppa generalized this idea using rational functions on an algebraic curve. After reviewing the basic definitions and results of the theory of algebraic curves in Section 4.2, we define the algebraic-geometric codes in Section 4.4. We give the application of modular curves to algebraic-geometric codes in Section 4.5. In order to produce a family of good algebraic-geometric codes, one needs a family of curves with a lot of rational points compared to the genus. The modular curves give an example of such a family of curves. Namely we show that there exist sequences of modular curve codes over finite fields $\mathbf{F}_q$ of increasing length, which are asymptotically better than the Gilbert-Varshamov bound for q being a square, $q \geq 49$ ([35]). In Chapter 5, we discuss a connection between binary linear codes and theta functions. One of the most important classes of codes is self-dual codes because many of the best known codes are of this type and they have a rich mathematical theory. After exhibiting a correspondence between self-dual binary codes and certain unimodular lattices, we show that the theta function associated to an even unimodular lattice in $\mathbf{R}^n$ is a modular form of weight $n/2$. In Section 5.2.2 and Section 5.3, we derive the MacWilliams identity for binary codes and Gleason's theorem from the transformation formula of the theta function. Finally we introduce the Leech lattice discovered by J. Leech in 1965. This lattice is an even unimodular lattice in $\mathbf{R}^{24}$ and does not contain any roots at all.

Appendix A[1] is devoted to the presentation of the fundamentals of the hyper-Kloosterman codes. We start with the definition of the hyper-Kloosterman sums. The codes obtained from the sums are called the hyper-Kloosterman codes. Using the theorems of Deligne-Katz and Deligne, we prove a result on the uniformity for the weight distribution of the hyper-Kloosterman codes. In Section A.7 we also give a divisibility theorem of the Hamming weights for the hyper-Kloosterman codes. A number of examples are given in Section A.6 and Section A.7.

As 'basso continuo' of this book, we will consider the weight distribution of codes. Much of the contents are devoted to finding the weight

---

[1]The contents of this appendix are based on the articles [2] and K. Chinen; On some properties of the hyper-Kloosterman codes, to appear in Tokyo J. Math.

distribution of linear codes. Recently, I. Duursma defined for a linear code its zeta function. The zeta function is a convenient tool to represent the weight distribution of algebraic-geometric codes.

Recently, a series of effective decoding algorithms for algebraic-geometric codes was worked out by several authors. Such algorithms decode up to half the minimum distance and have polynomial complexity. But, in this book, we do not describe the decoding problem for codes.

Finally, we would like to express our special appreciations to Dr. K. Chinen for applying his wonderful expertise of LaTeX to the final preparation of the manuscript.

Würzburg   Toyokazu Hiramatsu
July 2001   Günter Köhler

Chapter 1

LINEAR CODES

1. Coding Theory

In this chapter we introduce Coding Theory. This topic, known as the theory of error-correcting codes, is the study of methods for efficient and accurate transfer of information from one place to another. Coding theory deals with the problem of detecting and correcting transmission errors caused by noise on the channel. The following diagram provides a rough idea of a general information system:

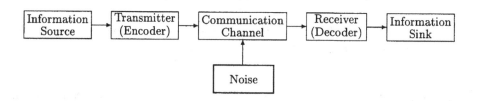

In coding theory one considers a set $\mathbf{F}$ of q distinct symbols which is called the *alphabet*. In practice q is generally 2 and $\mathbf{F} = \mathbf{F}_2$. In most of theory one takes $q = p^r$ (p prime) and $\mathbf{F} = \mathbf{F}_q$ the field of q elements. A *code* C with *word length* n is a subset of $\mathbf{F}^n$, an n-dimensional vector space over the finite field $\mathbf{F}$. Elements of $\mathbf{F}^n$ are called *words*, those of C *codewords*. We shall denote the number of elements of a finite set S by $|S|$ or by $\sharp S$.

REMARK 1.1 Any code over $\mathbf{F}_q$ is called a *q-ary code* (*binary* for $q = 2$, *ternary* for $q = 3$ and *quaternary* for $q = 2^2$).

In $\mathbf{F}^n$ we introduce so-called *Hamming distance* which is the natural distance function to use when one is interested in the number of errors in a word that is spelt incorrectly.

DEFINITION 1.1 For $x \in \mathbf{F}^n$ and $y \in \mathbf{F}^n$, we denote by $d(x, y)$ the number of coordinate positions in which x and y differ, that is

$$d(x, y) = |\{i \ : \ 1 \leq i \leq n; x_i \neq y_i\}|,$$

where $x = (x_1, \cdots, x_n)$ and $y = (y_1, \cdots, y_n)$.

DEFINITION 1.2 For $x \in \mathbf{F}^n$ we define the *weight* $w(x)$ of x by $w(x) = d(x, 0)$. (As usual 0 denotes the zero vector in $\mathbf{F}^n$.)

DEFINITION 1.3 The *minimum distance* $d(C)$ of a code C is the smallest Hamming distance between distinct codewords of C, namely

$$d(C) = d = \min\{d(x, y) \ : \ x \in C, \ y \in C, \ x \neq y\}.$$

If C has word length n, M words, and minimum distance d, then we call C an (n, M, d)-*code*.

PROPOSITION 1.1 The Hamming distance $d(x, y)$ is a distance function.

Proof. To see that the triangle inequality holds, let $x = (x_1, \cdots, x_n)$, $y = (y_1, \cdots, y_n)$ and $z = (z_1, \cdots, z_n)$. Then $d(x, z)$ is the number of positions in which x and z differ. If we denote the set of these positions by U, then

$$d(x, z) = |U| = |\{i \ : \ x_i \neq z_i\}|.$$

Let

$$S = \{i \ : \ x_i \neq z_i \text{ and } x_i = y_i\}$$

and

$$T = \{i \; : \; x_i \neq z_i \text{ and } x_i \neq y_i\}.$$

Then U is the disjoint union of S and T. Hence

$$d(x, z) = |S| + |T|.$$

It is immediate from the definitions of $d(x, y)$ and T that

$$|T| \leq d(x, y).$$

On the other hand if $i \in S$, then $y_i = x_i \neq z_i$. So

$$|S| \leq d(y, z),$$

and the triangle inequality follows. ∎

DEFINITION 1.4 For $\rho > 0$ and $x \in \mathbf{F}^n$ we define the *ball* of radius ρ with center at x by

$$B(x, \rho) = \{y \in \mathbf{F}^n \; : \; d(x, y) \leq \rho\}.$$

Consider a subset C of $\mathbf{F}^n$ with the property that any two distinct words of C have distance at least $2e + 1$. If we take any x in C and change t coordinates, where $t \leq e$ (that is, we make t errors), then the resulting word still resembles the original word more than it resembles any of the others. (That is, it has a smaller distance to x than to other words of C.) Therefore we can correct the t errors. Such a set C is called an *e-error-correcting code*. This definition implies that balls of radius e around distinct codewords are disjoint.

REMARK 1.2 The encoding and decoding operations can be described mathematically as functions:

$$E : \mathbf{F}^k \longrightarrow \mathbf{F}^n$$

and

$$D : \mathbf{F}^n \longrightarrow \mathbf{F}^k,$$

where $n > k$, E is one-to-one, and $D \circ E$ is the identity mapping on $\mathbf{F}^k$. Then $C = \operatorname{Im} E$ is the set of codewords, or just the code of length n over the alphabet $\mathbf{F}$. Words in $\mathbf{F}^n$ which are not in a ball $B(c, e)$ around a codeword $c \in C$ may be decoded incorrectly.

REMARK 1.3 A binary channel is *symmetric* if 0 and 1 are transmitted with equal accuracy; that is the probability of receiving the correct digit is independent of which digit, 0 or 1, is being transmitted. The reliability of a *binary symmetric channel (BSC)* is a real number p, $0 \leq p \leq 1$, where p is the probability that the digit sent is the digit received. In this case, $1 - p$ is the probability that the digit received is not the digit sent. The following diagram may clarify how a BSC operates:

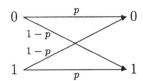

We will always assume that we are using a BSC with probability p satisfying $1/2 < p < 1$.

2.   Linear codes

A k-dimensional linear subspace C of $\mathbf{F}^n$ is called a *linear code*. We use the notation $[n, k]$-code if the code has dimension k and $[n, k, d]$-code if, moreover, the minimum distance is d. The error-correcting capacity of a code is determined by the minimum distance between all pairs of distinct codewords.

PROPOSITION 1.2 The minimum distance of a linear code is equal to the minimum weight among all nonzero codewords.

Proof. If $x \in C$ and $y \in C$, then $x - y \in C$ and

$$d(x, y) = d(x - y, 0) = w(x - y). \quad \blacksquare$$

We shall now look at two ways of describing a linear code C. The first is given by a *generator matrix* G. This is a matrix for which the rows are a set of basis vectors of the linear subspace C. This means that G is a k by n matrix over $\mathbf{F}$ and

$$C = \{aG \ : \ a \in \mathbf{F}^k\}.$$

We call two codes *equivalent* if one is obtained from the other by applying a fixed permutation to the positions for all codewords. The minimum distance of a code does not change under such a permutation. So we can assume that a generator matrix G of C has the so-called *standard form* $(I_k \, P)$, where I_k is the k by k identity matrix and P is k by $n - k$ matrix. A code C is called *systematic* if there is a k-subset of the coordinate positions such that to each possible k-tuple of entries in these positions there corresponds exactly one codeword. We have seen that every linear code is systematic.

Next we develop another matrix associated with a linear code C and closely connected with the generator matrix G.

DEFINITION 1.5 If C is a linear code of dimension k, then we define the *dual code* $C^\perp$ by

$$C^\perp = \{y \in \mathbf{F}^n \ : \ x \cdot y = 0 \text{ for all } x \in C\},$$

where $x \cdot y$ denotes the usual inner product in $\mathbf{F}^n$.

The code $C^\perp$ is a linear code of dimension $n - k$. If H is a generator matrix for $C^\perp$, then H is called a *parity check matrix* for C. Such a parity check matrix then defines the code C by

$$C = \{x \in \mathbf{F}^n \ : \ xH^\mathrm{T} = 0\}.$$

A code C is *self-orthogonal* provided that $C \subseteq C^\perp$, and *self-dual* provided that $C = C^\perp$.

PROPOSITION 1.3 If $G = (I_k \, P)$ is a generator matrix for C in standard form, then $H = (-P^T \, I_{n-k})$ is a parity check matrix for C.

Proof. This follows from $HG^T = -P^T + P^T = 0$. ∎

We observed that the minimum distance of a linear code is the minimum weight of any nonzero codeword. The minimum distance of a linear code can also be determined from a parity check matrix for the code.

THEOREM 1.4 Let H be a parity check matrix for a linear code C. Then C has minimum distance d if and only if any set of $d-1$ columns of H is linearly independent, and at least one set of d columns of H is linearly dependent.

Proof. We denote the columns of H by $H_1, \cdots, H_n$. Note that for $c = (c_1, \cdots, c_n)$,
$$Hc^T = c_1 H_1 + \cdots c_n H_n.$$
Suppose that there is a nonzero codeword $c = (c_1, \cdots, c_n)$ of weight d. Since $c \in C$ we have $Hc^T = 0$ and hence
$$Hc^T = c_1 H_1 + \cdots c_n H_n = 0.$$
This is a non-trivial linear relation among the d columns of H corresponding to the positions with nonzero entries in C. Hence there exist d columns of H which are linearly dependent. It is also clear now that any $d-1$ columns of H are linearly independent; otherwise, there would be a nonzero codeword c with weight $w(c) \leq d-1$.

Now suppose, conversely, that H has d columns which are linearly dependent and that any $d-1$ columns of H are linearly independent. Then we see, as above, that there is no nonzero codeword c with weight $w(c) \leq d-1$ and that $w(c) = d$ for some codeword c. Therefore, by Proposition 1.2, the code C has minimum distance d. ∎

THEOREM 1.5 (THE SINGLETON BOUND) For any $[n, k, d]$-linear code, $d \leq n - k + 1$.

Proof. Recall that from Theorem 1.4 the parity check matrix H of an $[n, k, d]$-linear code is an n by $n-k$ matrix such that every $d-1$ columns of H are linearly independent. On the other hand $n-k$ is the rank of H and is the maximum number of linearly independent columns of H. Hence $d-1 \leq n-k$ or, equivalently, $d \leq n-k+1$. ∎

REMARK 1.4 One of the most important parameters of a code C is its so-called *information rate* R defined by

$$R = n^{-1} \log_q |C|.$$

For a linear code of dimension k, we have $R = k/n$. Good codes are ones for which $R = k/n$ is not too small, but for which d is also not too small. These conditions are clearly somewhat incompatible or trade off (Theorem 1.5). And while it is known by Shannon's Theorem (A Mathematical Theory of Communication, 1948) that there exist codes with R nearly 1 for which the probability of decoding a random received word incorrectly is arbitrarily small, it may be necessary to take n very large (hence large k as well) to achieve this. Hence identifying good codes is a delicate balancing act, and much effort has been devoted both to finding explicit good codes, and to developing theoretical bounds on the parameters of codes.

3. Cyclic codes

Many of the most interesting codes that we shall study are cyclic. We define these as follows.

DEFINITION 1.6 A linear code C is called *cyclic* if for each codeword $(c_0, c_1, \cdots, c_{n-1})$ in C, $(c_{n-1}, c_0, c_1, \cdots, c_{n-2})$ is also a codeword of C.

PROPOSITION 1.6 If C is cyclic then so is $C^{\perp}$.

Proof. Let C be a cyclic $[n, k]$-code. We define the cyclic shift of a word $c = (c_0, c_1, \cdots, c_{n-1})$ by $c^1 = (c_{n-1}, c_0, c_1, \cdots, c_{n-2})$. We will show that $h^1 \in C^\perp$ whenever $h \in C^\perp$, thus proving the cyclicity of $C^\perp$. For every $c \in C$ we have

$$
\begin{aligned}
h^1 \cdot c &= h_{n-1}c_0 + h_0c_1 + \cdots + h_{n-2}c_{n-1} \\
&= h_0c_1 + \cdots + h_{n-2}c_{n-1} + h_{n-1}c_0 \\
&= h \cdot c^{n-1} = 0
\end{aligned}
$$

because $h \in C^\perp$ and $c^{n-1} = (c_1, c_2, \cdots c_{n-1}, c_0) \in C$. Hence $h^1 \in C^\perp$ as required. ∎

The aim of this section is to exhibit a connection between cyclic codes and polynomial rings. In this section we make the restriction $\gcd(n, q) = 1$. Let $R = \mathbf{F}_q[x]$ be the ring of all polynomials with coefficients in $\mathbf{F}_q$, and let S be the ideal generated by $x^n - 1$. Then the residue class ring R/S is isomorphic to $\mathbf{F}_q{}^n$ considered as an additive group (or as a $\mathbf{F}_q$-vector space). An isomorphism is given by

$$
c = (c_0, c_1 \cdots, c_{n-1}) \longleftrightarrow c_0 + c_1x + \cdots c_{n-1}x^{n-1} = c(x).
$$

Note that distinct polynomials of degree $< n$ represent distinct residue classes. So we can say that $c(x)$ belongs to R/S. From now on we do not distinguish between words and polynomials of degree $< n$ mod $(x^n - 1)$. Multiplication by x in this ring amounts to a *cyclic shift* of the words. From this it follows that a cyclic code C corresponds to an ideal I in R/S:

THEOREM 1.7 A set I of polynomials in the ring R/S represents a cyclic code C if and only if I is an ideal of R/S.

Proof. Let I be an ideal and a_1, a_2 be elements of I. Then by the ideal property, $a_1 + a_2 \in I$ and $\lambda a_1 \in I$ for any element λ of R/S. Taking $\lambda \in \mathbf{F}_q$, we see that C is linear. Taking $\lambda = x$, we conclude that C is closed under the cyclic shift. Hence C is a cyclic code.

Conversely, let C be cyclic. We have to show that I has the ideal property. Let c be any codeword, represented by $c(x)$ in I, and let $p(x) = p_0 + p_1x + \cdots + p_{n-1}x^{n-1}$ be any polynomial in R/S. Then

$p(x)c(x) = p_0 c(x) + p_1 x c(x) + \cdots p_{n-1} x^{n-1} c(x)$ is a combination of cyclic shifts of c, so it must be in C. Hence $p(x)c(x) \in I$. The remaining items are easy to see. ∎

Since the ring R/S is a principal ideal ring, every cyclic code in $\mathbf{F}^n$ is a principal ideal generated by some polynomial $g(x)$ that divides $x^n - 1$. We shall call $g(x)$ the *generator of the cyclic code*.

This fact is explained as follows. Let I be an ideal different from $\{0\}$ in R/S. Then the preimage of I with respect to the canonical map is an ideal in R. Clearly, this preimage is a principal ideal containing S, and hence it is generated by a polynomial $g(x)$ which divides $x^n - 1$.

Let

$$g(x) = g_0 + g_1 x + \cdots + g_{n-k} x^{n-k}$$

be the generator of a cyclic code C. Then the dimension of C is k and the words $g(x), xg(x), \cdots, x^{k-1} g(x)$ form a basis of the code C:

THEOREM 1.8 Every cyclic $[n, k]$-code C has a generator matrix G of the form

$$G = \begin{pmatrix} g_0 & g_1 & \cdots & g_{n-k} & 0 & \cdots & \cdots & 0 \\ 0 & g_0 & g_1 & \cdots & g_{n-k} & 0 & \cdots & 0 \\ \vdots & & \cdots & & & \cdots & & \vdots \\ \vdots & & & \cdots & & & \cdots & 0 \\ 0 & \cdots & \cdots & 0 & g_0 & g_1 & \cdots & g_{n-k} \end{pmatrix}$$

in which each row is a cyclic shift of the previous row, and neither g_0 nor g_{n-k} is zero.

Proof. Clearly we can assume that $g_{n-k} \neq 0$. Since C is cyclic and $g(x)$ has minimal degree, we obtain that $g_0 \neq 0$. Therefore, k is the rank of G. We show that $\{g(x), xg(x), \cdots, x^{k-1} g(x)\}$ is a basis for C: let $c(x)$ be a polynomial in $(g(x))$, the ideal of R/S generated by $g(x)$.

Then $c(x) = f(x)g(x)$ in R/S for some polynomial $f(x)$, and hence $c(x) = f(x)g(x) + d(x)(x^n - 1)$ in $\mathbf{F}[x]$ for some polynomial $d(x)$. Since $g(x)|(x^n - 1)$,

$$c(x) = \left(f(x) + \frac{(x^n - 1)d(x)}{g(x)} \right) g(x)$$

in $\mathbf{F}[x]$. Hence every element of $(g(x))$ is of the form $a(x)g(x)$ with $a(x) \in \mathbf{F}[x]$ and $\deg a(x) < k$. ∎

Given the generator polynomial $g(x)$ of a cyclic $[n, k]$-code, the code can be put into systematic form. Suppose that the information to be encoded is $u = (u_0, u_1, \cdots, u_{k-1})$. The corresponding polynomial is

$$u(x) = u_0 + u_1 x + \cdots + u_{k-1} x^{k-1}.$$

Multiplying $u(x)$ by x^{n-k}, we obtain a polynomial of degree $n - 1$ or less,

$$x^{n-k}u(x) = u_0 x^{n-k} + u_1 x^{n-k+1} + \cdots + u_{k-1} x^{n-1}.$$

Dividing $x^{n-k}u(x)$ by the generator polynomial $g(x)$, we have

$$x^{n-k}u(x) = a(x)g(x) + b(x),$$

where the degree of $b(x)$ must be $n - k - 1$ or less, and we put

$$-b(x) = b_0 + b_1 x + \cdots + b_{n-k-1} x^{n-k-1}.$$

Then the polynomial $-b(x) + x^{n-k}u(x)$ is a code polynomial of the cyclic code generated by $g(x)$, and we have

$$\begin{aligned} -b(x) + x^{n-k}u(x) = \ & b_0 + b_1 x + \cdots b_{n-k-1} x^{n-k-1} \\ & + u_0 x^{n-k} + u_1 x^{n-k+1} + \cdots + u_{k-1} x^{n-1}, \end{aligned}$$

which corresponds to the codeword

$$(b_0, b_1, \cdots, b_{n-k-1}, u_0, u_1, \cdots, u_{k-1}).$$

REMARK 1.5 From these results we see that each ideal of R/S corresponds to a unique divisor of $x^n - 1$, its generator, so the number of cyclic codes of length n over $\mathbf{F}$ is the number of monic divisors of $x^n - 1$. The assumption $\gcd(n, q) = 1$ guarantees that $x^n - 1$ does not have multiple divisors in $\mathbf{F}[x]$. The factorization of $x^n - 1$ into irreducibles over $\mathbf{F}_2$ is given in the following table.

n	factorization
3	$(1+x)(1+x+x^2)$
5	$(1+x)(1+x+x^2+x^3+x^4)$
7	$(1+x)(1+x+x^3)(1+x^2+x^3)$
9	$(1+x)(1+x+x^2)(1+x^3+x^6)$
11	$(1+x)(1+x+\cdots+x^{10})$
13	$(1+x)(1+x+\cdots+x^{12})$
15	$(1+x)(1+x+x^2)(1+x+x^2+x^3+x^4)(1+x+x^4)(1+x^3+x^4)$
17	$(1+x)(1+x+x^2+x^4+x^6+x^7+x^8)(1+x^3+x^4+x^5+x^8)$
19	$(1+x)(1+x+\cdots+x^{18})$
21	$(1+x)(1+x+x^2)(1+x^2+x^3)(1+x+x^3)(1+x^2+x^4+x^5+x^6)$
	$\times(1+x+x^2+x^4+x^6)$
23	$(1+x)(1+x+x^5+x^6+x^7+x^9+x^{11})(1+x^2+x^4+x^5+x^6+x^{10}+x^{11})$
25	$(1+x)(1+x+x^2+x^3+x^4)(1+x^5+x^{10}+x^{15}+x^{20})$
63	$(1+x)(1+x+x^2)(1+x^2+x^3)(1+x+x^3)(1+x+x^6)(1+x^3+x^6)$
	$\times(1+x^5+x^6)(1+x+x^2+x^4+x^6)(1+x+x^3+x^4+x^6)$
	$\times(1+x+x^2+x^5+x^6)(1+x+x^4+x^5+x^6)(1+x^2+x^3+x^5+x^6)$
	$\times(1+x^2+x^4+x^5+x^6)$

Next we give a polynomial interpretation of the parity check matrix of a cyclic code. Let $C = (g(x))$ be a cyclic $[n, k]$-code. Hence $g(x)|(x^n - 1)$ so $g(x)h(x) = x^n - 1$ for some $h(x)$. Now $\deg g(x) = n - k$ so $\deg h(x) = k$, and $h(x)$ is monic. The polynomial $h(x)$ is called the *check polynomial* of C. In spite of its name $h(x)$ does not in general generate $C^\perp$ but there is a close connection between $h(x)$ and $C^\perp$ as we shall see below.

PROPOSITION 1.9 A polynomial $c(x)$ corresponds to a codeword of C if and only if $c(x)h(x) \equiv 0 \bmod (x^n - 1)$.

Proof. Let c be an element of C. Then $c(x) \equiv a(x)g(x) \bmod (x^n - 1)$ for some $a(x) \in \mathbf{F}[x]$. Hence $c(x)h(x) \equiv a(x)g(x)h(x) \equiv 0 \bmod (x^n - 1)$. Conversely, suppose $c(x)h(x) \equiv 0 \bmod (x^n - 1)$. Divide $c(x)$ by $g(x)$ to get $c(x) = g(x)q(x) + r(x)$ with $\deg r(x) < n-k$. Then $r(x)h(x) \equiv 0 \bmod (x^n - 1)$, namely, $r(x)h(x)$ is a multiple of $x^n - 1$. But $\deg r(x) < n-k$ and $\deg h(x) = k$, so $\deg r(x)h(x) < n$, and hence $r(x)h(x) = 0$. Therefore $r(x) = 0$ so $c(x) = g(x)q(x)$; that is $c \in C$. ∎

We have seen that the duals of cyclic codes are themselves cyclic. It remains to find the generator polynomial of the dual, which will then give us a generator matrix for $C^\perp$, in other words a parity check matrix for C.

THEOREM 1.10 If $h(x)$ is the check polynomial of the cyclic $[n, k]$-code C, then $\overline{h}(x) = h_0^{-1}x^k h(x^{-1})$ is the generator polynomial of $C^\perp$, where h_0 denotes the constant term of $h(x)$.

Proof. Since $h_0 \neq 0$, $\overline{h}(x)$ is well-defined and monic of degree k. From $g(x)h(x) = x^n - 1$, we get $g(x^{-1})h(x^{-1}) = x^{-n} - 1$, and this can be written as $-x^n h_0 h_0^{-1} h(x^{-1})g(x^{-1}) = x^n - 1$, i.e., $-h_0 g(x^{-1})x^{n-k}\overline{h}(x) = x^n - 1$. Now $x^{n-k}g(x^{-1})$ is a polynomial of degree $n - k$ since $g_0 \neq 0$. Hence $\overline{h}(x)$ is a monic divisor of $x^n - 1$, with degree k, so $(\overline{h}(x))$ is a cyclic $[n, n - k]$-code. But $\dim C^\perp = n - k$. So if we can show that $(\overline{h}(x)) \subseteq C^\perp$, it will follow that $(\overline{h}(x)) = C^\perp$.

The polynomials $\overline{h}(x)$ and $g(x)$ represent the words $(h_k, h_{k-1}, \cdots, h_0, 0, \cdots, 0)$ and $(g_0, g_1, \cdots, g_{n-k}, 0, \cdots, 0)$, respectively. These are orthogonal because their inner product $g_0 h_k + g_1 h_{k-1} + \cdots$ is simply the coefficient of x^k in $h(x)g(x) = x^n - 1$, and $0 < k < n$. Similarly the inner products of $\overline{h}(x)$ with $xg(x)$, $\cdots$, $x^{k-1}g(x)$ are the coefficients of x^{k-1}, $\cdots$, x in $x^n - 1$, respectively, all of which are zero. Hence $\overline{h}(x) \in C^\perp$ because it is orthogonal to every row of the generator matrix for C. But $C^\perp$ is cyclic, so every cyclic shift of $\overline{h}(x)$ is in $C^\perp$. Every polynomial multiple of $\overline{h}(x)$ corresponds to a linear combination of cyclic shifts of $\overline{h}(x)$, so by linearity these are all in $C^\perp$, too. That is $(\overline{h}(x)) \subseteq C^\perp$, as claimed. ∎

COROLLARY 1.11 If $h(x) = h_0 + h_1 x + \cdots + h_k x^k$ is the check polynomial of C, then the matrix

$$H = \begin{pmatrix} 0 & \cdots & \cdots & 0 & h_k & \cdots & \cdots & h_1 & h_0 \\ 0 & \cdots & 0 & h_k & \cdots & \cdots & h_1 & h_0 & 0 \\ \vdots & & \cdots & & & & & & \vdots \\ 0 & \cdots & & & & & \cdots & & \vdots \\ h_k & \cdots & \cdots & h_1 & h_0 & 0 & \cdots & \cdots & 0 \end{pmatrix}$$

is a generator matrix for $C^{\perp}$, that is, a parity check matrix for C.

4. Finite fields

In this section we summarize without proof some facts from the theory of finite fields.

Given any power $q = p^r$ of a prime p, there exists up to isomorphism exactly one finite field of order q which is denoted by $\mathbf{F}_q$. The number p is called the *characteristic* of $\mathbf{F}_q$ and is denoted by $\mathrm{Char}\,(\mathbf{F}_q) = p$. The field $\mathbf{F}_q$ can be constructed as the splitting field of the polynomial $x^q - x$ over $\mathbf{F}_p$. It can be shown that the set of all roots of this polynomial forms a field which is thus equal to $\mathbf{F}_q$. Then it follows that $\mathbf{F}_q$ is isomorphic to the residue class ring $\mathbf{F}_p[x]/(g(x))$ for some irreducible polynomial $g(x)$ with degree r in $\mathbf{F}_p[x]$.

The multiplicative group $\mathbf{F}_q^{\times}$ of nonzero elements of $\mathbf{F}_q$ is cyclic, and a generator of $\mathbf{F}_q^{\times}$ is called a *primitive element* of $\mathbf{F}_q$.

We define the *trace function* $\mathrm{Tr}_{K/F} : K \to F$ on $K = \mathbf{F}_{q^m}$ over $F = \mathbf{F}_q$ by

$$\mathrm{Tr}_{K/F}(\alpha) = \mathrm{Tr}\,(\alpha) = \alpha + \alpha^q + \cdots + \alpha^{q^{m-1}}$$

for $\alpha \in K$. The trace function $\mathrm{Tr}_{K/F}$ has the following properties:

(1) Tr is not identically zero;

(2) $\mathrm{Tr}_{K/F}(\alpha + \beta) = \mathrm{Tr}_{K/F}(\alpha) + \mathrm{Tr}_{K/F}(\beta)$ for all $\alpha, \beta \in K$;

(3) $\text{Tr}_{K/F}(a\alpha) = a\text{Tr}_{K/F}(\alpha)$ for all $\alpha \in K$ and $a \in F$;

(4) Tr is a nontrivial linear transformation from K onto F, where both K and F are viewed as linear spaces over F;

(5) $\text{Tr}_{K/F}(\alpha) = ma$ for all $\alpha \in F$;

(6) $\text{Tr}_{K/F}(\alpha^q) = \text{Tr}_{K/F}(\alpha)$ for all $\alpha \in K$;

(7) if L is a finite extension of K, then

$$\text{Tr}_{L/F}(\alpha) = \text{Tr}_{K/F}(\text{Tr}_{L/K}(\alpha))$$

for all $\alpha \in L$.

The trace of a vector $c = (c_1, \cdots, c_n) \in K^n$ is defined by

$$\text{Tr}(c) = (\text{Tr}(c_1), \cdots, \text{Tr}(c_n)).$$

5. BCH codes

In this section we introduce a very important class of codes known as BCH codes. They are named after R. C. Bose, D. K. Ray-Chaudhuri, and A. Hocquenghem.

DEFINITION 1.7 Let δ be an integer with $2 \le \delta \le n$. A cyclic code C of length n over $\mathbf{F}_q$ is called a *BCH code* of designed distance δ if its generator $g(x)$ is the least common multiple of the minimal polynomials of β^l, β^{l+1}, $\cdots$, $\beta^{l+\delta-2}$ for some l,

$$g(x) = \text{lcm}\,\{M_{\beta^l}(x), \cdots, M_{\beta^{l+\delta-2}}(x)\},$$

where β is a primitive n-th root of unity in some extension field of $\mathbf{F}_q$, and $M_{\beta^i}(x)$ is the minimal polynomial of β^i over $\mathbf{F}_q$. If $l = 1$, C is called a BCH code in the *narrow sense*. If $n = q^m - 1$, that is, if β is a primitive element of $\mathbf{F}_{q^m}$, then C is called *primitive*. Usually we take $l = 1$.

The terminology "designed distance" is explained by the following result.

THEOREM 1.12 Let C be an $[n, k]$-BCH code over $\mathbf{F}_q$ of designed distance δ. Then the minimum distance of C is at least δ.

Proof. We define the $\delta - 1$ by n matrix H over some extension field of $\mathbf{F}_q$ by

$$H = \begin{pmatrix} 1 & \beta^l & \beta^{2l} & \cdots & \beta^{(n-1)l} \\ 1 & \beta^{l+1} & \beta^{2(l+1)} & \cdots & \beta^{(n-1)(l+1)} \\ \cdots & \cdots & & & \cdots \\ 1 & \beta^{l+\delta-2} & \beta^{2(l+\delta-2)} & \cdots & \beta^{(n-1)(l+\delta-2)} \end{pmatrix}.$$

It is easy to see that a word c is in the code C if and only if $Hc^{\mathrm{T}} = 0$. If $c \in C$, then $c(x) = a(x)g(x)$ for some polynomial $a(x) \in \mathbf{F}_q[x]$. Thus $c(x)$ is a multiple of $M_{\beta^j}(x)$ for $l \leq j \leq l + \delta - 2$. It follows that

$$c_0 + c_1\beta^j + c_2\beta^{2j} + \cdots + c_{n-1}\beta^{(n-1)j} = c(\beta^j) = a(\beta^j)g(\beta^j) = 0$$

for $l \leq j \leq l + \delta - 2$. This yields $Hc^{\mathrm{T}} = 0$. Conversely, from $Hc^{\mathrm{T}} = 0$ it follows that $c(\beta^j) = 0$ for $l \leq j \leq l + \delta - 2$. Thus $c(x)$ is a common multiple of $M_{\beta^j}(x)$ for $l \leq j \leq l + \delta - 2$. Then it is also a multiple of their least common multiple $g(x)$, so $c(x) \in (g(x))$ and $c \in C$. Any $\delta - 1$ columns of H form, after extracting powers of β, a Vandermonde matrix. Since this matrix has determinant $\neq 0$, the columns are linearly independent. Now it follows from Theorem 1.4 that c cannot have minimum distance less than δ. ∎

DEFINITION 1.8 A *Reed-Solomon code*, abbreviated *RS-code* over $\mathbf{F}_q$ is a BCH code of length $n = q - 1$.

In this case, β is a primitive element of $\mathbf{F}_q$ itself, and the minimal polynomial of β^i over $\mathbf{F}_q$ is simply $M_{\beta^i}(x) = x - \beta^i$. So in the narrow sense case, the generator of an RS-code has the form $g(x) = \prod_{i=1}^{\delta-1}(x - \beta^i)$, where β is a primitive element of $\mathbf{F}_q$. By Theorem 1.12 this code has minimum distance at least δ, and by the Singleton bound (Theorem 1.5), the distance cannot be larger. Therefore, RS-codes are $[n, n - d + 1, d]$-codes.

EXAMPLE 1.1 Let $\alpha \in \mathbf{F}_{16}$ be a root of $x^4 + x + 1 \in \mathbf{F}_2[x]$, then α and α^3 have the minimal polynomials $M_\alpha(x) = x^4 + x + 1$ and $M_{\alpha^3}(x) = x^4 + x^3 + x^2 + x + 1$ over $\mathbf{F}_2$, respectively. Both $M_\alpha(x)$ and $M_{\alpha^3}(x)$ are divisors of $x^{15} - 1$. Both α^2 and α^4 have $M_\alpha(x)$ as their minimal polynomial. Hence we can define a binary BCH code C with designed distance $\delta = 5$ and with generator polynomial $g(x) = M_\alpha(x) \cdot M_{\alpha^3}(x)$. Since g divides $f \in \mathbf{F}_2[x]/(x^{15} - 1)$ if and only if $f(\alpha) = f(\alpha^3) = 0$, we see that

$$H = \begin{pmatrix} 1 & \alpha & \alpha^2 & \cdots & \alpha^{14} \\ 1 & \alpha^3 & \alpha^6 & \cdots & \alpha^{42} \end{pmatrix}$$

is a parity check matrix for C over $\mathbf{F}_{2^4}$. By Theorem 1.12, the minimum distance d of C is at least 5. In this case, d is just 5, since $g(x)$ is a codeword of weight 5.

Chapter 2

DIOPHANTINE EQUATIONS AND CYCLIC CODES

1.   Diagonal equations over finite fields

We consider a polynomial equation of the type

$$a_1 x_1^{d_1} + \cdots + a_s x_s^{d_s} = b, \qquad (2.1)$$

where s (≥ 2), $d_1, \cdots, d_s$ are positive integers, $b \in \mathbf{F}_q$, $a_1, \cdots, a_s \in \mathbf{F}_q^{\times}$, and $q = p^r$ with a prime p. Such an equation is called a *diagonal equation*. By the number N of solutions of this equation (2.1) in $\mathbf{F}_q{}^s$ we mean the number of s-tuples $(\gamma_1, \cdots, \gamma_s) \in \mathbf{F}_q{}^s$ for which $a_1 \gamma_1^{d_1} + \cdots + a_s \gamma_s^{d_s} = b$.

THEOREM 2.1 Let χ_a be the additive character of $\mathbf{F}_q$ defined by $\chi_a(x) = \exp((2\pi i/p)\mathrm{Tr}\,(ax))$ for $a \in \mathbf{F}_q$. Suppose that $d_1 = \cdots = d_s = d$. Then the number N of solutions of the equation (2.1) is

$$N = q^{-1} \sum_{a \in \mathbf{F}_q} \chi_a(-b) \prod_{i=1}^{s} S_{aa_i},$$

where $S_u = \sum_{x \in \mathbf{F}_q} \chi_u(x^d)$.

Proof. We put

$$F(x_1, \cdots, x_s) = a_1 x_1^d + \cdots + a_s x_s^d - b,$$

17

$$S = \sum_{x=(x_1,\cdots,x_s)\in \mathbf{F}_q^s} \sum_{a\in\mathbf{F}_q} \chi_a(F(x_1,\cdots,x_s)).$$

Because of a well-known character sum property,

$$\sum_{a\in\mathbf{F}_q} \chi_a(F(x_1,\cdots,x_s)) = \begin{cases} q, & \text{if } F(x_1,\cdots,x_s) = 0, \\ 0, & \text{otherwise.} \end{cases}$$

Now by interchanging the two summations it follows that

$$qN = S = \sum_{a\in\mathbf{F}_q} \sum_{(x_1,\cdots,x_s)\in\mathbf{F}_q{}^s} \chi_a(F(x_1,\cdots,x_s)).$$

By the morphism property of χ_a, the inner sum is equal to the product of $\chi_a(-b)$ by

$$\sum_{(x_1,\cdots,x_s)\in\mathbf{F}_q{}^s} \left(\prod_{i=1}^{s} \chi_a(a_i x_i^d)\right)$$

which is also equal to

$$\prod_{i=1}^{s} \left(\sum_{x\in\mathbf{F}_q} \chi_a(a_i x^d)\right). \quad \blacksquare$$

COROLLARY 2.2 The number N of solutions of the diagonal equation

$$x_1^d + \cdots + x_s^d = 0$$

is given by

$$N = p^{-r} \sum_{a\in\mathbf{F}_q} (S_a)^s,$$

where $S_a = \sum_{x\in\mathbf{F}_q} \omega^{\mathrm{Tr}(ax^d)}$, $\omega = \exp(2\pi i/p)$.

2.   The number of solutions and weight distributions of cyclic codes

First we recall some facts from the theory of cyclic codes.

Let $m(x)$ be an irreducible monic divisor of $x^n - 1$ over $\mathbf{F}_p$. Then $(m(x))$ is a maximal ideal, and the cyclic code generated by

$(x^n - 1)/m(x)$ is called an *irreducible cyclic code*. We have $m(x) = M_\alpha(x)$, where α is an n-th root of unity and $M_\alpha(x)$ is the minimal polynomial of α over $\mathbf{F}_p$. We denote by $C(\alpha)$ such a code. The code $C(\alpha)$ called *non-degenerate* if α is a primitive n-th root of unity. We suppose that $C(\alpha)$ is non-degenerate. Let L be the splitting field of $x^n - 1$ over $\mathbf{F}_p$. It is known that $C(\alpha)$ is the image of L by the one-to-one mapping μ from L into $\mathbf{F}_p{}^n$ which is defined by

$$\mu(c) = (\mathrm{Tr}\,(c), \mathrm{Tr}\,(c\alpha), \cdots, \mathrm{Tr}\,(c\alpha^{n-1})),$$
$$= \sum_{i=0}^{n-1} \mathrm{Tr}\,(c\alpha^i)x^i$$

where Tr denotes the trace of L over $\mathbf{F}_p$.

PROPOSITION 2.3 Let a be in $\mathbf{F}_q$, $q = p^r$, and suppose that $p^r - 1 = nd$ and $p - 1$ divides n. Then the sum S_a in Corollary 2.2 is given by

$$S_a = p^r - \frac{p}{p-1}dw(\mu(c)),$$

where $c = \tau(a)$, τ is the trace of $\mathbf{F}_q$ over the splitting field L of $x^n - 1$ over $\mathbf{F}_p$, and $w(\mu(c))$ is the weight of the codeword $\mu(c)$ in the non-degenerate code $C(\alpha)$.

Proof. We consider the equation

$$\mathrm{Tr}_{\mathbf{F}_q/\mathbf{F}_p}(ax^d) = \lambda \tag{2.2}$$

for $\lambda \in \mathbf{F}_p$. Let N_λ be the number of its solutions x in $\mathbf{F}_q$. If $\lambda \neq 0$, then the equation (2.2) is equivalent to

$$\mathrm{Tr}\,(\lambda^{-1}ax^d) = 1. \tag{2.3}$$

From $(p-1)|n$ it follows that $\mathbf{F}_p^\times$ is contained in the multiplicative subgroup G of order n of $\mathbf{F}_q^\times$. The elements of G are the d-th powers in $\mathbf{F}_q^\times$. Therefore, there exists some x_λ in $\mathbf{F}_q^\times$ such that $\lambda^{-1} = x_\lambda^d$. We put $y = x_\lambda x$; then the equation (2.3) becomes

$$\mathrm{Tr}\,(ay^d) = 1.$$

Hence $N_\lambda = N_1$ for $\lambda \neq 0$. That is,

$$S_a = \sum_{\lambda \in \mathbf{F}_p} N_\lambda \omega^\lambda = N_0 + N_1 \sum_{\lambda \neq 0} \omega^\lambda$$

$$= N_0 + N_1 \left(-1 + \sum_{\lambda=0}^{p-1} \omega^\lambda \right).$$

Because of $\sum_{\lambda=0}^{p-1} \omega^\lambda = 0$, we have

$$S_a = N_0 - N_1.$$

On the other hand, let τ be the trace of $\mathbf{F}_q$ over L, and we put $c = \tau(a)$. By the transitivity property of the trace functions ((7) in 1.4), we have

$$\begin{aligned}
\mathrm{Tr}_{\mathbf{F}_q/\mathbf{F}_p}(a\alpha^i) &= \mathrm{Tr}_{L/\mathbf{F}_p}(\mathrm{Tr}_{\mathbf{F}_q/L}(a\alpha^i)) \\
&= \mathrm{Tr}_{L/\mathbf{F}_p}(\tau(a\alpha^i)) \\
&= \mathrm{Tr}_{L/\mathbf{F}_p}(\alpha^i \tau(a)) \\
&= \mathrm{Tr}_{L/\mathbf{F}_p}(c\alpha^i).
\end{aligned}$$

Hence

$$n - w(\mu(c)) = \left| \{ i \; : \; 0 \leq i \leq n-1, \mathrm{Tr}_{\mathbf{F}_q/\mathbf{F}_p}(a\alpha^i) = 0 \} \right|.$$

Because of $nd = p^r - 1$, the equation $x^d = \alpha^i$ has exactly d solutions in $\mathbf{F}_q^\times$. Therefore $N_0 = 1 + d(n - w(\mu(c)))$. Furthermore

$$|\mathbf{F}_q| = \sum_{\lambda \in \mathbf{F}_p} N_\lambda,$$

which means that $p^r = N_0 + (p-1)N_1$. By combining the above results, we have

$$S_a = p^r - \frac{p}{p-1} dw(\mu(c)). \quad \blacksquare$$

Let C be a linear code of length n, and let A_i denote the number of codewords in C which are of weight i. We call the polynomial

$$\sum_{i=0}^{n} A_i x^{n-i} y^i$$

the *weight enumerator* of C and denote it by $W_C(x, y)$. This is a homogeneous polynomial of degree n in the variables x and y. Observe that we can rewrite this polynomial as

$$W_C(x, y) = \sum_{u \in C} x^{n-w(u)} y^{w(u)}.$$

The sequence $\{A_i\}_{i=0}^n$ is called the *weight distribution* of C.

THEOREM 2.4 (WOLFMANN [38], [39]) With notations and assumptions as in Proposition 2.3, let N be the number of solutions $(x_1, \cdots, x_s)$ of $x_1^d + \cdots + x_s^d = 0$ in $\mathbf{F}_q{}^s$. Then

$$N = \frac{p^{s-\nu}}{(p-1)^s} \sum_{i=0}^n A_i \{(p-1)p^{r-1} - di\}^s, \tag{2.4}$$

where A_i is the number of codewords in $C = C(\alpha)$ of weight i, and ν is the multiplicative order of p modulo n.

REMARK 2.1 From the assumptions $p^r - 1 = nd$ and $(p-1)|n$ in Proposition 2.3 it follows that $d \mid \{(p^r - 1)/(p - 1)\}$.

Proof. By Corollary 2.2 and Proposition 2.3, N is given by

$$N = \frac{p^{s-r}}{(p-1)^s} \sum_{a \in \mathbf{F}_q} \left((p-1)p^{r-1} - dw(\mu(\tau(a)))\right)^s.$$

The multiplicative order ν of p modulo n is a divisor of r, and the splitting field L of $x^n - 1$ over $\mathbf{F}_p$ is $\mathbf{F}_{p^\nu}$. For every $c \in L$, the number of elements $a \in \mathbf{F}_q$ such that $c = \tau(a)$ is $(p^\nu)^{r/\nu-1} = p^{r-\nu}$. Therefore we have

$$\sum_{a \in \mathbf{F}_q} \left((p-1)p^{r-1} - dw(\mu(\tau(a)))\right)^s$$
$$= p^{r-\nu} \sum_{c \in L} \left((p-1)p^{r-1} - dw(\mu(c))\right)^s.$$

Hence

$$N = \frac{p^{s-\nu}}{(p-1)^s} \sum_{c \in L} \left((p-1)p^{r-1} - dw(\mu(c))\right)^s. \tag{2.5}$$

The mapping μ is one-to-one and thus the number of c in L such that $w(\mu(c)) = i$ is the number A_i of codewords in $C(\alpha)$ of weight i. Applying this to (2.5), we have the formula (2.4). ∎

EXAMPLE 2.1 Let $p = 2$, $d = 7$, $r = 6$ and $n = 9$. Then the number N of solutions of $x_1^7 + \cdots + x_s^7 = 0$ over $\mathbf{F}_{2^6}$ is given by

$$N = 2^{2s-6}\{16^s + 9^{s+1} + 27 \cdot 2^s + 27(-5)^s\}.$$

Proof. The irreducible divisor $x^6 + x^3 + 1$ of $x^9 - 1$ over $\mathbf{F}_2$ is a minimal polynomial of α over $\mathbf{F}_2$, where α is a primitive 9-th root of unity over $\mathbf{F}_2$. Therefore, the generator of $C(\alpha)$ is given by $g(x) = (x^9 - 1)/M_\alpha(x) = x^3 + 1$. Hence, the generator matrix of $C(\alpha)$ is

$$G = \begin{pmatrix} 1 & 0 & 0 & 1 & 0 & 0 & 0 & 0 & 0 \\ 0 & 1 & 0 & 0 & 1 & 0 & 0 & 0 & 0 \\ 0 & 0 & 1 & 0 & 0 & 1 & 0 & 0 & 0 \\ 0 & 0 & 0 & 1 & 0 & 0 & 1 & 0 & 0 \\ 0 & 0 & 0 & 0 & 1 & 0 & 0 & 1 & 0 \\ 0 & 0 & 0 & 0 & 0 & 1 & 0 & 0 & 1 \end{pmatrix}.$$

The codewords of $C(\alpha)$ are the linear combinations over $\mathbf{F}_2$ of the rows of G. Thus the weight distribution of $C(\alpha)$ is given by

$$A_0 = 1, A_2 = 9, A_4 = 27, A_6 = 27,$$

and $A_i = 0$ for all other i. From Theorem 2.4, we have the expected result. ∎

Theorem 2.4 shows that, under the condition that d divides $(p^r - 1)/(p - 1)$, we are able to explicitly calculate the number of solutions of $x_1^d + \cdots + x_s^d = 0$ if we know the weight distribution of the associated irreducible cyclic code.

Chapter 3

ELLIPTIC CURVES, HECKE OPERATORS AND WEIGHT DISTRIBUTION OF CODES

1. Elliptic curves over finite fields

Every elliptic curve is given by an equation of the form

$$E : y^2 + a_1 xy + a_3 y = x^3 + a_2 x^2 + a_4 x + a_6. \tag{3.1}$$

Such an equation is called a *Weierstrass equation* for E. If the coefficients a_i can all be chosen from some field K then we say that E is defined over the field K. The indexing of the coefficients appears to be strange. It is explained as follows. We assign the weights 3, 2 and i to x, y and a_i, respectively. Then every term in the equation has total weight 6. If $\text{Char}\,(K) \neq 2$, then the change of variables

$$(x, y) \mapsto \left(x,\; y - \frac{a_1}{2}x - \frac{a_3}{2} \right)$$

transforms E to the curve

$$E' : y^2 = x^3 + b_2 x^2 + b_4 x + b_6.$$

If $\text{Char}\,(K) \neq 2, 3$, then the change of variables

$$(x, y) \mapsto \left(\frac{x - 3b_2}{36},\; \frac{y}{216} \right)$$

further transforms E' to the curve

$$E'' : y^2 = x^3 + ax + b.$$

23

Let E be a curve given by a Weierstrass equation (3.1). Define the quantities

$$d_2 = a_1^2 + 4a_2, \quad d_4 = 2a_4 + a_1 a_3, \quad d_6 = a_3^2 + 4a_6,$$

$$d_8 = a_1^2 a_6 + 4a_2 a_6 - a_1 a_3 a_4 + a_2 a_3^2 - a_4^2,$$

$$c_4 = d_2^2 - 24d_4,$$

$$\Delta = -d_2^2 d_8 - 8d_4^3 - 27d_6^2 + 9d_2 d_4 d_6,$$

$$j(E) = c_4^3 / \Delta.$$

The quantity Δ is called the *discriminant* of E, and $j(E)$ is called the *j-invariant* of E if $\Delta \neq 0$. The condition $\Delta \neq 0$ is equivalent to the non-singularity of E.

A technical definition of non-singularity will be given in Section 4.2.1. Geometrically it means that the curve has a well-defined tangent line everywhere.

Let K be a field of characteristic 2. Then we have $j(E) = a_1^{12}/\Delta$. If $j(E) \neq 0$, then the change of variables

$$(x, y) \mapsto \left(a_1^2 x + \frac{a_3}{a_1}, \ a_1^3 y + \frac{a_1^2 a_4 + a_3^2}{a_1^3} \right)$$

transforms E to the curve

$$E_1 : y^2 + xy = x^3 + a_2 x^2 + a_6.$$

For E_1 we have $\Delta = a_6$ and $j(E) = 1/a_6$. If $j(E) = 0$, then the change of variables

$$(x, y) \mapsto (x + a_2, \ y)$$

transforms E to the curve

$$E_2 : y^2 + a_3 y = x^3 + a_4 x + a_6.$$

For E_2 we have $\Delta = a_3^4$ and $j(E_2) = 0$.

1.1   The group law

Let E be an elliptic curve given by the Weierstrass equation (3.1), defined over the field K. By a point on E we mean either a solution of the equation (3.1) with coordinates in some extension field L over K, or the unique point on E at infinity in the projective plane over K. This point has homogeneous coordinates $(0:1:0)$ and will be denoted by O. (See Section 4.2 for a more detailed explanation and generalizations.) The points on E form an abelian group $(E, +)$ under an addition $+$ which can be defined as follows.

(1) The neutral element is the point O at infinity.

(2) If $P = (x_1, y_1) \neq O$ then $-P = (x_1, -y_1 - a_1 x_1 - a_3)$.

(3) If $P \neq O$, $Q \neq O$, $P \neq -Q$, then $P + Q = -R$, where R is the third point of intersection of either the line $\overline{PQ}$ if $P \neq Q$, or the tangent line to the curve at P if $P = Q$, with the curve.

Let E be defined over K, and let L be an extension field of K. The set of L-*rational points* of E, denoted by $E(L)$, is the set of points of E both of whose coordinates lie in L, together with the point O. Then $E(L)$ is a subgroup of $(E, +)$.

Explicit formulae for the coordinates of $P + Q$ in terms of the coordinates of P and Q are easy to derive. Let $P = (x_1, y_1)$, $Q = (x_2, y_2)$ and $P + Q = (x_3, y_3)$. Let l be the line passing through P and Q if $P \neq Q$, or the tangent line to the curve at P in the case $P = Q$. The slope of l is

$$\lambda = \begin{cases} \dfrac{y_2 - y_1}{x_2 - x_1} & \text{if } P \neq Q, \\[2mm] \dfrac{3x_1^2 + 2a_2 x_1 + a_4 - a_1 y_1}{2y_1 + a_1 x_1 + a_3} & \text{if } P = Q. \end{cases}$$

If $\beta = y_1 - \lambda x_1$, then the equation defining l is $y = \lambda x + \beta$. The third point of intersection of l with the curve is easy to calculate:

$$x_3 = \lambda^2 + a_1 \lambda - a_2 - x_1 - x_2$$

and

$$y_3 = -(\lambda + a_1)x_3 - \beta - a_3.$$

REMARK 3.1 (1) Addition formula in the case of Char $(K) \neq 2, 3$. If $P = (x_1, y_1) \in E''$, then $-P = (x_1, -y_1)$. If $Q = (x_2, y_2) \in E''$, $Q \neq -P$, then $P + Q = (x_3, y_3)$, where

$$x_3 = \lambda^2 - x_1 - x_2, \quad y_3 = \lambda(x_1 - x_3) - y_1,$$

and

$$\lambda = \begin{cases} \dfrac{y_2 - y_1}{x_2 - x_1} & \text{if } P \neq Q, \\[2mm] \dfrac{3x_1^2 + a}{2y_1} & \text{if } P = Q. \end{cases}$$

(2) Addition formula in the case of Char $(K) = 2$. When $j(E) \neq 0$, let $P = (x_1, y_1) \in E_1$. Then $-P = (x_1, y_1 + x_1)$. If $Q = (x_2, y_2) \in E_1$, $Q \neq -P$, then $P + Q = (x_3, y_3)$, where

$$x_3 = \begin{cases} \left(\dfrac{y_1 + y_2}{x_1 + x_2}\right)^2 + \dfrac{y_1 + y_2}{x_1 + x_2} + x_1 + x_2 + a_2 & \text{if } P \neq Q, \\[4mm] x_1^2 + \dfrac{a_6}{x_1^2} & \text{if } P = Q, \end{cases}$$

and

$$y_3 = \begin{cases} \left(\dfrac{y_1 + y_2}{x_1 + x_2}\right)(x_1 + x_3) + x_3 + y_1 & \text{if } P \neq Q, \\[4mm] x_1^2 + \left(x_1 + \dfrac{y_1}{x_1}\right)x_3 + x_3 & \text{if } P = Q. \end{cases}$$

When $j(E) = 0$, let $P = (x_1, y_1) \in E_2$. Then $-P = (x_1, y_1 + a_3)$. If $Q = (x_2, y_2) \in E_2$, $Q \neq -P$, then $P + Q = (x_3, y_3)$, where

$$x_3 = \begin{cases} \left(\dfrac{y_1 + y_2}{x_1 + x_2}\right)^2 + x_1 + x_2 & \text{if } P \neq Q, \\[4mm] \dfrac{x_1^4 + a_4^2}{a_3^2} & \text{if } P = Q, \end{cases}$$

and

$$y_3 = \begin{cases} \left(\dfrac{y_1 + y_2}{x_1 + x_2}\right)(x_1 + x_3) + y_1 + a_3 & \text{if } P \neq Q, \\[4mm] \left(\dfrac{x_1^2 + a_4}{a_3}\right)(x_1 + x_3) + y_1 + a_3 & \text{if } P = Q. \end{cases}$$

EXAMPLE 3.1 We start with the two points $P = (-2, 3)$, $Q = (2, 5)$ on the elliptic curve

$$E : y^2 = x^3 + 17$$

over $\mathbf{Q}$. The line connecting P and Q has slope $1/2$, so its equation is

$$y = \frac{1}{2}x + 4.$$

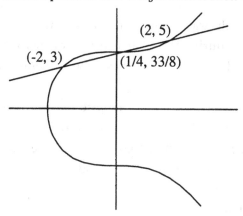

Figure 3.1

Substituting this into the equation for E gives

$$x^3 - \frac{1}{4}x^2 - 4x + 1 = 0.$$

This must have $x = -2$ and $x = 2$ as roots, so it factors as

$$(x - 2)(x + 2)\left(x - \frac{1}{4}\right) = 0.$$

Therefore, by taking the line through the two known solutions $(-2, 3)$, $(2, 5)$, we have found the rational solution $(1/4, 33/8)$ for our elliptic curve E. This procedure is illustrated in Figure 3.1.

If (x, y) is a point on the elliptic curve E, then the point $(x, -y)$ will also be a point on E. This is clear from the symmetry of E about the x-axis. Therefore we have $P + Q = (1/4, -33/8)$.

1.2   Hasse's theorem

Let $q = p^r$, where p is the characteristic of $\mathbf{F}_q$. Let E be an elliptic curve with Weierstrass equation (3.1) defined over $\mathbf{F}_q$. For the numbers $|E(\mathbf{F}_q)|$ of points in $E(\mathbf{F}_q)$ we have the following result.

THEOREM 3.1 (HASSE) Let $|E(\mathbf{F}_q)| = q + 1 - t$. Then $|t| \leq 2\sqrt{q}$.

This beautiful result was conjectured by E. Artin in the 1920's and proven by H. Hasse during the 1930's. A generalized version was proven by A. Weil in the 1940's, and this was again generalized by P. Deligne in the 1970's.

It is known that there exists an elliptic curve E over $\mathbf{F}_q$ such that $E(\mathbf{F}_q)$ has order $q + 1 - t$ with $t^2 \leq 4q$ if and only if one of the following conditions is satisfied: 1. $t \not\equiv 0 \pmod{p}$, 2. r is even, and $t = \pm 2\sqrt{q}$, $t = \pm\sqrt{q}$ and $p \not\equiv 1 \pmod{3}$ or $t \equiv 0 \pmod{p}$ and $p \not\equiv 1 \pmod{4}$, 3. r is odd, and $t = 0$ or $t = \pm\sqrt{pq}$ and $p = 2$ or 3 (Waterhouse, Ann. Ecole Norm. Sup. 2 (1969)).

EXAMPLE 3.2 (GAUSS) Let N_p be the number of solutions in $\mathbf{F}_p \times \mathbf{F}_p$ of $y^2 = x^3 - 432$, $p \neq 2, 3$. Then $N_p = p$ for $p \equiv 2 \pmod{3}$. If $p \equiv 1 \pmod{3}$, there are integers A and B, unique up to sign, such that $4p = A^2 + 27B^2$. If the sign of A is so chosen that $A \equiv 1 \pmod{3}$, then $N_p = p + A - 2$. In particular, $|N_p - p| \leq 2\sqrt{p}$.

REMARK 3.2 Yu. I. Manin gave a completely elementary proof of Theorem 3.1.

REMARK 3.3 The *zeta function* of an elliptic curve E over $\mathbf{F}_q$ is defined by the formal Dirichlet series

$$Z(E, s) = \exp\left(\sum_{n=1}^{\infty} |E(\mathbf{F}_{q^n})| \frac{q^{-ns}}{n}\right).$$

Then we have

$$Z(E, s) = \frac{1 - aq^{-s} + q^{1-2s}}{(1 - q^{-s})(1 - q^{1-s})},$$

where $a = q + 1 - |E(\mathbf{F}_q)|$, and $1 - aT + qT^2 = (1 - \alpha T)(1 - \beta T)$ with $|\alpha| = |\beta| = \sqrt{q}$. Therefore, the zeta function $Z(E, s)$ has poles at $s = 0$ and $s = 1$. Also, Theorem 3.1 is equivalent to the assertion that the roots of $1 - aT + qT^2$ are complex conjugates of each other. These roots have absolute value equal to $1/\sqrt{q}$. This condition is also equivalent to the assertion that the zeros of the zeta function $Z(E, s)$ are all on the line Re $s = 1/2$.

2.   Modular forms and Hecke operators

In this section we collect without proof some facts from the theory of modular forms that we shall need in this course. For proofs we refer to readily available textbook, for example, that by J. P. Serre [31]. We begin with some basic notations.

2.1 $SL_2(\mathbf{Z})$ and its congruence subgroups

We put

$$\Gamma = SL_2(\mathbf{Z}) = \left\{ \begin{pmatrix} a & b \\ c & d \end{pmatrix} : a, b, c, d \in \mathbf{Z}, ad - bc = 1 \right\}.$$

The group Γ is called the *full modular group*. Let N be a positive integer. The *principal congruence subgroup of level N* is denoted by $\Gamma(N)$ and consists of all matrices in Γ satisfying

$$\begin{pmatrix} a & b \\ c & d \end{pmatrix} \equiv \begin{pmatrix} 1 & 0 \\ 0 & 1 \end{pmatrix} \pmod{N}.$$

Since this is the kernel of the natural mapping (reduction mod N)

$$SL_2(\mathbf{Z}) \longrightarrow SL_2(\mathbf{Z}/N\mathbf{Z}),$$

$\Gamma(N)$ is a normal subgroup of finite index in $SL_2(\mathbf{Z})$. The *Hecke subgroup of level N* is denoted by $\Gamma_0(N)$ and consists of all matrices

$$\begin{pmatrix} a & b \\ c & d \end{pmatrix} \in \Gamma$$

such that $N|c$. Since

$$\Gamma(N) \subseteq \Gamma_0(N) \subseteq \Gamma,$$

$\Gamma_0(N)$ has finite index in Γ. The group $\Gamma_1(N)$ consists of all matrices $\gamma \in \Gamma$ satisfying

$$\gamma \equiv \begin{pmatrix} 1 & * \\ 0 & 1 \end{pmatrix} \pmod{N}.$$

Clearly

$$\Gamma(N) \subseteq \Gamma_1(N) \subseteq \Gamma_0(N) \subseteq \Gamma.$$

A *congruence subgroup* of Γ is a subgroup which contains $\Gamma(N)$ for some N. Thus $\Gamma_0(N)$, $\Gamma_1(N)$ are examples of congruence subgroups.

2.2   The upper half-plane

Let $\mathfrak{H}$ denote the *upper half-plane*

$$\mathfrak{H} = \{z = x + iy \ : \ x, y \in \mathbf{R}, y > 0\}.$$

Let $GL_2^+(\mathbf{R})$ be the group of 2 by 2 matrices with real entries and positive determinant. Then $GL_2^+(\mathbf{R})$ acts on $\mathfrak{H}$ as a group of holomorphic automorphisms by

$$\gamma : z \mapsto \frac{az+b}{cz+d} \quad \text{for } \gamma = \begin{pmatrix} a & b \\ c & d \end{pmatrix} \in GL_2^+(\mathbf{R}).$$

Let $\mathfrak{H}^*$ denote the union of $\mathfrak{H}$ and the rational numbers $\mathbf{Q}$ together with a symbol ∞. The action of Γ on $\mathfrak{H}$ can be extended to $\mathfrak{H}^*$ by defining

$$\begin{pmatrix} a & b \\ c & d \end{pmatrix} \cdot \infty = \frac{a}{c} \quad (c \neq 0),$$

$$\begin{pmatrix} a & b \\ 0 & d \end{pmatrix} \cdot \infty = \infty,$$

and

$$\begin{pmatrix} a & b \\ c & d \end{pmatrix} \cdot \frac{r}{s} = \frac{ar+bs}{cr+ds}$$

for rational numbers r/s with $\gcd(r, s) = 1$, with the understanding that when $cr + ds = 0$, the right hand side of the above equation is ∞. The rational numbers together with ∞ are called *cusps*.

If G is a discrete subgroup of $SL_2(\mathbf{R})$, then the orbit space $\mathfrak{H}^*/G$ can be given the structure of a compact Riemann surface X_G. We will be interested in the case that G is a congruence subgroup of Γ. In that case, the algebraic curve corresponding to X_G is called a *modular curve*. In the case $G = \Gamma(N)$, $\Gamma_1(N)$ or $\Gamma_0(N)$, the corresponding modular curve is denoted by $X(N)$, $X_1(N)$, or $X_0(N)$, respectively.

2.3   Modular forms and cusp forms

Let f be a holomorphic function on $\mathfrak{H}$ and k a positive integer. For

$$\gamma = \begin{pmatrix} a & b \\ c & d \end{pmatrix} \in GL_2^+(\mathbf{R}),$$

define

$$(f|_k\gamma)(z) = (\det\gamma)^{\frac{k}{2}}(cz+d)^{-k}f\left(\frac{az+b}{cz+d}\right).$$

For fixed k, the mapping $\gamma : f \mapsto f|_k\gamma$ defines an action of $GL_2^+(\mathbf{R})$ on the space of holomorphic functions on $\mathfrak{H}$. Let G be a subgroup of finite index in Γ. Let f be a holomorphic function on $\mathfrak{H}$ such that $f|_k\gamma = f$ for all $\gamma \in G$. Since G has finite index in Γ,

$$\begin{pmatrix} 1 & 1 \\ 0 & 1 \end{pmatrix}^M = \begin{pmatrix} 1 & M \\ 0 & 1 \end{pmatrix} \in G$$

for some positive integer M. Hence $f(z+M) = f(z)$ for all $z \in \mathfrak{H}$. So, f has a Fourier expansion at infinity,

$$f(z) = \sum_{n=-\infty}^{\infty} a_n q_M{}^n \quad \text{with } q_M = e^{\frac{2\pi i z}{M}}.$$

We say that f is *holomorphic at infinity* if $a_n = 0$ for all $n < 0$. We say it *vanishes at infinity* if $a_n = 0$ for all $n \leq 0$. Let $\sigma \in \Gamma$. Then $\sigma^{-1}G\sigma$ also has finite index in Γ and $(f|_k\sigma)|_k\gamma = f|_k\sigma$ for all $\gamma \in \sigma^{-1}G\sigma$. So for any $\sigma \in \Gamma$, $f|_k\sigma$ also has a Fourier expansion at infinity. We say that f is *holomorphic at the cusps* if $f|_k\sigma$ is holomorphic at infinity for all $\sigma \in \Gamma$.

Let N be a positive integer and χ a Dirichlet character mod N. A *modular form* on $\Gamma_0(N)$ of type (k, χ) is a holomorphic function f on $\mathfrak{H}$ such that

(1) $f|_k \begin{pmatrix} a & b \\ c & d \end{pmatrix} = \chi(d)f$

for all $\begin{pmatrix} a & b \\ c & d \end{pmatrix} \in \Gamma_0(N)$, and

(2) f is holomorphic at the cusps.

Note that (1) implies $f|_k\gamma = f$ for all $\gamma \in \Gamma_1(N)$. The Fourier expansion of such a form f is

$$f(z) = \sum_{n=0}^{\infty} a_n q^n, \quad q = e^{2\pi i z}.$$

The integer k is called the *weight* of f. Such a modular form is called a *cusp form* if it vanishes at the cusps. The modular forms on $\Gamma_0(N)$ of type (k, χ) form a complex linear space $M_k(\Gamma_0(N), \chi)$, and this has as a subspace the set $S_k(\Gamma_0(N), \chi)$ of all cusp forms. The subspace has a canonical complement,

$$M_k(\Gamma_0(N), \chi) = \mathcal{E}_k(\Gamma_0(N), \chi) \oplus S_k(\Gamma_0(N), \chi)$$

and the space $\mathcal{E}_k$ is called the space spanned by *Eisenstein series*. These spaces are finite dimensional.

2.4   Hecke operators

Let p denote a prime number and $f(z) = \sum_{n=0}^{\infty} a_n q^n$ be a modular form on $\Gamma_0(N)$ of type (k, χ). The *Hecke operators* T_p and U_p are defined by

$$
f|_k T_p(z) = \sum_{n=0}^{\infty} a_{np} q^n + \chi(p) p^{k-1} \sum_{n=0}^{\infty} a_n q^{np} \quad \text{if } p \nmid N,
$$

$$
f|_k U_p(z) = \sum_{n=0}^{\infty} a_{np} q^n \quad \text{if } p \mid N.
$$

It is easy to show that $f|T_p$ and $f|U_p$ are also modular forms on $\Gamma_0(N)$ of type (k, χ), and they are cusp forms if f is a cusp form.

THEOREM 3.2 (HECKE-PETERSSON) The operators T_p for $p \nmid N$ are commuting linear transformations of $S_k(\Gamma_0(N), \chi)$. The space can be decomposed as a direct sum of common eigenspaces of the operators T_p.

Let $f \in S_k(\Gamma_0(N), \chi)$. We say that f is an *eigenform* if f is an eigenfunction for all the Hecke operators T_p. If

$$f(z) = \sum_{n=1}^{\infty} a_n e^{2\pi i n z} \tag{3.2}$$

is the Fourier expansion at ∞, and $a_1 = 1$, we call it *normalized*. For the Fourier expansion (3.2), we attach an L-function by

$$L(s, f) = \sum_{n=1}^{\infty} \frac{a_n}{n^s}.$$

Then we have:

THEOREM 3.3 (HECKE-PETERSSON) The space $S_k(\Gamma_0(N), \chi)$ has a basis of normalized eigenfunctions for all operators T_p. If f is a normalized newform, its Dirichlet series $L(s, f)$ extends to an entire function and has an Euler product expansion

$$L(s, f) = \prod_{p|N} \left(1 - \frac{a_p}{p^s}\right)^{-1} \prod_{p \nmid N} \left(1 - \frac{a_p}{p^s} + \frac{\chi(p)}{p^{2s+1-k}}\right)^{-1}$$

which converges absolutely for $\mathrm{Re}\, s > (k+2)/2$.

REMARK 3.4 Suppose $N'|N$ and χ is a Dirichlet character modulo N'. If $f(z) \in S_k(\Gamma_0(N'), \chi)$ and $dN'|N$, then $f(dz) \in S_k(\Gamma_0(N), \chi)$. The forms on $\Gamma_0(N)$ which may be obtained in this way from a divisor N' of N ($N' \neq N$), span a subspace of $S_k(\Gamma_0(N), \chi)$ called the space of *oldforms*. Its canonical complement is denoted $S_k^{\mathrm{new}}(\Gamma_0(N), \chi)$ and the eigenforms in this space are called *newforms*.

REMARK 3.5 The space $S_k(\Gamma_1(N), 1)$ can be decomposed according to the Dirichlet characters $\chi \bmod N$ which are the characters of $\Gamma_0(N)/\Gamma_1(N)$:

$$S_k(\Gamma_1(N)) = S_k(\Gamma_1(N), 1) = \bigoplus_\chi S_k(\Gamma_0(N), \chi).$$

The Hecke operators on $S_k(\Gamma_1(N))$ respect the above decomposition of this space. We have a unified definition of the Hecke operators T_m for all positive integers m. For $f(z) = \sum_{n=1}^\infty a_n e^{2\pi i n z}$ in $S_k(\Gamma_0(N), \chi)$, the action of T_m is defined by

$$f|_k T_m(z) = \sum_{n=1}^\infty b_n q^n \quad (q = e^{2\pi i z}),$$

$$b_n = \sum_{d|\gcd(m,n)} \chi(d) d^{k-1} a_{mn/d^2},$$

where we put $\chi(d) = 0$ whenever $\gcd(d, N) > 1$.

The *Eichler-Selberg trace formula* gives an expression for the *traces of Hecke operators* in terms of class numbers of binary quadratic forms (H. Cohen [3]):

THEOREM 3.4 Let $k \geq 2$ be an integer for which $\chi(-1) = (-1)^k$, and let N_χ be the conductor of χ. For every integer $n \geq 1$, the trace of the Hecke operator T_n acting on the space $S_k(\Gamma_0(N), \chi)$ of cusp forms is given by

$$\mathrm{Tr}\,(T_n) = A_1 + A_2 + A_3 + A_4,$$

where the terms on the right hand side are defined as follows.

(1) $$A_1 = n^{\frac{k}{2}-1}\chi(\sqrt{n}) \cdot \frac{k-1}{12}\psi(N),$$

where $\chi(\sqrt{n}) = 0$ whenever $\sqrt{n} \notin \mathbf{Z}$ and

$$\psi(N) = N\prod_{p|N}\left(1 + \frac{1}{p}\right).$$

(2) $$A_2 = -\frac{1}{2}\sum_{\substack{t\in\mathbf{Z}\\ t^2<4n}}\frac{\rho^{k-1} - \bar{\rho}^{k-1}}{\rho - \bar{\rho}}\sum_f h_w\left(\frac{t^2 - 4n}{f^2}\right)\mu(t, f, n).$$

Here ρ and $\bar{\rho}$ denote the roots of the polynomial $x^2 - tx + n$, in the inner sum f runs over the positive divisors of $t^2 - 4n$ for which $(t^2 - 4n)/f^2 \in \mathbf{Z}$ is congruent to 0 or 1 mod 4, and $h_w(\Delta)$ is given by

$$h_w(-3) = \frac{1}{3}, \quad h_w(-4) = \frac{1}{2},$$
$$h_w(\Delta) = h(\Delta) \quad \text{for } \Delta < -4,$$

where $h(\Delta)$ denotes the class number of Δ (*cf.* §3.2 of this chapter). The numbers $\mu(t, f, n)$ are given by

$$\mu(t, f, n) = \frac{\psi(N)}{\psi(N/N_f)}\sum_{\substack{x \ (\mathrm{mod}\ N)\\ x^2 - tx + n \equiv 0 \ (\mathrm{mod}\ N_f N)}}\chi(x),$$

where $N_f = \gcd(N, f)$.

(3) $$A_3 = -\sum_{\substack{d|n\\ 0<d\leq\sqrt{n}}}{}' d^{k-1}\sum_{\substack{c|N\\ (c,N/c)|(N/N_\chi, n/d-d)}}\phi\left(\gcd\left(c, \frac{N}{c}\right)\right)\chi(y).$$

Here ϕ denotes the Euler function and the prime in the first summation indicates that the contribution of the term $d = \sqrt{n}$, if it occurs, should be multiplied by $1/2$. The number y is defined modulo $N/(c, N/c)$ by $y \equiv d \pmod{c}$, $y \equiv n/d \pmod{N/c}$.

$$(4) \qquad A_4 = \begin{cases} \displaystyle\sum_{\substack{0 < t \mid n \\ (N, n/t) = 1}} t, & \text{if } k = 2, \chi = \text{id.,} \\ \\ 0, & \text{otherwise.} \end{cases}$$

3. Weight distribution of linear codes

In this section we study the weight distributions of the Melas codes. These codes are related to some deep results concerning elliptic curves over finite fields and modular forms. We start with an explanation of the MacWilliams identities and Delsarte's theorem, relating the weight distribution of a code and that of its dual.

3.1 The MacWilliams identities and Delsarte's theorem

Let C be a linear code of length n, and let A_i denote the number of codewords in C which are of weight i. In Section 2.2 we called the polynomial

$$W_C(x, y) = \sum_{i=0}^{n} A_i x^{n-i} y^i$$

the weight enumerator of C. By setting $x = 1$, we have a perfectly good weight enumerator in the variable y:

$$W_C(1, y) = W_C(y) = \sum_{i=0}^{n} A_i y^i.$$

The following theorem is a remarkable *formula of MacWilliams*, which enables the weight enumerator of any linear code C to be obtained from the weight enumerator of its dual code $C^\perp$.

THEOREM 3.5 (THE MACWILLIAMS IDENTITY FOR BINARY LINEAR CODES) If C is a binary $[n, k]$-code with dual code $C^{\perp}$, then

$$W_{C^{\perp}}(z) = 2^{-k} \sum_{i=0}^{n} A_i (1 - z)^i (1 + z)^{n-i}.$$

The following three lemmas are required only for the proof of Theorem 3.5.

LEMMA 3.6 Let C be a binary linear $[n, k]$-code and suppose y is a vector in $\mathbf{F}_2^n$ which is not in $C^{\perp}$. Then the inner product $x \cdot y$ is equal to 0 and 1 equally often as x runs over the codewords of C.

Proof. Let $A = \{x \in C \ : \ x \cdot y = 0\}$ and $B = \{x \in C \ : \ x \cdot y = 1\}$. Because of $y \notin C^{\perp}$ there exists a codeword $u \in B$. Then

$$u + A \subseteq B;$$

for, if $x \in A$, then $(u + x) \cdot y = u \cdot y + x \cdot y = 1 + 0 = 1$. Similarly,

$$u + B \subseteq A.$$

Hence, $|A| = |u + A| \leq |B| = |u + B| \leq |A|$. Therefore $|A| = |B|$.

Another proof of Lemma 3.6: We consider the linear form $C \rightarrow \mathbf{F}_2$ which is given by $x \mapsto x \cdot y$. Its kernel A is different from C because of $y \notin C^{\perp}$. Thus $C/A \simeq \mathbf{F}_2$. Therefore, A and its complement have the same numbers of elements. ∎

LEMMA 3.7 Let C be a binary linear $[n, k]$-code and let y be any vector of $\mathbf{F}_2^n$. Then

$$\sum_{x \in C} (-1)^{x \cdot y} = \begin{cases} 2^k, & \text{if } y \in C^{\perp}, \\ 0, & \text{if } y \notin C^{\perp}. \end{cases}$$

Proof. If $y \in C^{\perp}$, then $x \cdot y = 0$ for all $x \in C$, and so

$$\sum_{x \in C} (-1)^{x \cdot y} = |C| = 2^k.$$

If $y \notin C^{\perp}$, then by Lemma 3.6, as x runs over the elements of C, $(-1)^{x \cdot y}$ is equal to 1 and -1 equally often, giving

$$\sum_{x \in C} (-1)^{x \cdot y} = 0. \quad \blacksquare$$

LEMMA 3.8 For any fixed x in $\mathbf{F}_2^n$ we have the identity

$$\sum_{y \in \mathbf{F}_2^n} z^{w(y)} (-1)^{x \cdot y} = (1 - z)^{w(x)} (1 + z)^{n - w(x)}.$$

Proof. We obtain

$$\begin{aligned}
\sum_{y \in \mathbf{F}_2^n} z^{w(y)} (-1)^{x \cdot y} &= \sum_{y_1 = 0}^{1} \cdots \sum_{y_n = 0}^{1} z^{y_1 + \cdots + y_n} (-1)^{x_1 y_1 + \cdots + x_n y_n} \\
&= \sum_{y_1 = 0}^{1} \cdots \sum_{y_n = 0}^{1} \left(\prod_{i=1}^{n} z^{y_i} (-1)^{x_i y_i} \right) \\
&= \prod_{i=1}^{n} \left(\sum_{j=0}^{1} z^j (-1)^{j x_i} \right) \\
&= (1 - z)^{w(x)} (1 + z)^{n - w(x)},
\end{aligned}$$

since

$$\sum_{j=0}^{1} z^j (-1)^{j x_i} = \begin{cases} 1 + z, & \text{if } x_i = 0, \\ 1 - z, & \text{if } x_i = 1. \end{cases} \quad \blacksquare$$

Proof of Theorem 3.5. We express the polynomial

$$f(z) = \sum_{x \in C} \left(\sum_{y \in \mathbf{F}_2^n} (-1)^{x \cdot y} z^{w(y)} \right)$$

in two ways. On the one hand, using Lemma 3.8,

$$\begin{aligned}
f(z) &= \sum_{x \in C} (1 - z)^{w(x)} (1 + z)^{n - w(x)} \\
&= \sum_{i=0}^{n} A_i (1 - z)^i (1 + z)^{n-i}.
\end{aligned}$$

On the other hand, reversing the order of summation and using Lemma 3.7,

$$
\begin{aligned}
f(z) &= \sum_{y \in \mathbf{F}_2^n} z^{w(y)} \left(\sum_{x \in C} (-1)^{x \cdot y} \right) \\
&= \sum_{y \in C^\perp} z^{w(y)} \cdot 2^k \\
&= 2^k W_{C^\perp}(z). \quad \blacksquare
\end{aligned}
$$

Let C be any linear $[n, k]$-code over $\mathbf{F}_{q^m}$. The *subfield subcode* $\mathrm{Res}\,(C)$ of C with respect to $\mathbf{F}_q$ is the set of codewords in C each of whose components is in $\mathbf{F}_q$. This is a linear code of length n and dimension $\leq k$ over $\mathbf{F}_q$. The *trace code* of C over $\mathbf{F}_q$ is defined by

$$
\mathrm{Tr}\,(C) = \{\mathrm{Tr}\,(c) \ : \ c \in C\}.
$$

The trace code of C is also a linear code over $\mathbf{F}_q$. The following theorem of Delsarte exhibits a dual relation between subfield subcodes and trace codes:

THEOREM 3.9 (DELSARTE) Let C be a linear code of length n over $\mathbf{F}_{q^m}$. Then we have
$$
(\mathrm{Res}\,(C))^\perp = \mathrm{Tr}\,(C^\perp).
$$

Proof. We first show that $(\mathrm{Res}\,(C))^\perp \supseteq \mathrm{Tr}\,(C^\perp)$. Let $c = (c_1, \cdots, c_n)$ be in $C^\perp$ and $b = (b_1, \cdots, b_n)$ in $\mathrm{Res}\,(C)$. Then by the linearity of the trace function,

$$
\begin{aligned}
(\mathrm{Tr}\,(c)) \cdot b &= \sum_{i=1}^n (\mathrm{Tr}\,(c_i)) b_i \\
&= \mathrm{Tr} \left(\sum_{i=1}^n c_i b_i \right) = \mathrm{Tr}\,(0) = 0.
\end{aligned}
$$

Hence, $\mathrm{Tr}\,(c) \in (\mathrm{Res}\,(C))^\perp$. Therefore, $\mathrm{Tr}\,(C^\perp) \subseteq (\mathrm{Res}\,(C))^\perp$.

Next we show that $(\mathrm{Tr}\,(C^\perp))^\perp \subseteq \mathrm{Res}\,(C)$. Let $a = (a_1, \cdots, a_n) \in (\mathrm{Tr}\,(C^\perp))^\perp$. Since $a_i \in \mathbf{F}_q$ for $1 \leq i \leq n$, we need only show that $a \in C$.

Let $b = (b_1, \cdots, b_n)$ be a vector in $C^{\perp}$. Then $\beta b \in C^{\perp}$ for all $\beta \in \mathbf{F}_{q^m}$, and

$$0 = a \cdot (\mathrm{Tr}\,(\beta b)) = \sum_{i=1}^{n} a_i(\mathrm{Tr}\,(\beta b_i))$$

$$= \mathrm{Tr}\,\left(\beta \sum_{i=1}^{n} a_i b_i\right) = \mathrm{Tr}\,(\beta(a \cdot b)).$$

Hence $a \cdot b = 0$, and $a \in C$. Therefore $(\mathrm{Tr}\,(C^{\perp}))^{\perp} \subseteq \mathrm{Res}\,(C)$. It follows that $\mathrm{Tr}\,(C^{\perp}) \supseteq (\mathrm{Res}\,(C))^{\perp}$. ∎

EXAMPLE 3.3 Let $\mathbf{F}_4 = \{0, 1, a, a^2\}$, $n = 2$, $k = 1$,

$$C = \mathrm{span}\left\{\begin{pmatrix} 1 \\ a \end{pmatrix}\right\} = \left\{\begin{pmatrix} 0 \\ 0 \end{pmatrix}, \begin{pmatrix} 1 \\ a \end{pmatrix}, \begin{pmatrix} a \\ a^2 \end{pmatrix}, \begin{pmatrix} a^2 \\ 1 \end{pmatrix}\right\}.$$

For $q^m = 2^2$, $q = 2$, we obtain

$$\mathrm{Res}\,(C) = C \cap \mathbf{F}_2^2 = \left\{\begin{pmatrix} 0 \\ 0 \end{pmatrix}\right\}.$$

Thus $k = \dim C = 1$, but $\dim \mathrm{Res}\,(C) = 0$. Delsarte's theorem is satisfied with $((\mathrm{Res}\,(C))^{\perp} = \mathrm{Tr}\,(C^{\perp}) = \mathbf{F}_2^2$.

3.2 The weight distribution of the Melas codes

Let $q = 2^m > 4$. Let α be a primitive element of $\mathbf{F}_q$. Let $m(x) \in \mathbf{F}_2[x]$ be the minimal polynomial of α and $m_-(x) \in \mathbf{F}_2[x]$ that of α^{-1}. The ideal in the ring $\mathbf{F}_2[x]/(x^{q-1} - 1)$ which is generated by $m(x)m_-(x)$ is called a *Melas code*, and it is denoted by $M(q)$. The definition of the Melas codes is very similar to that of BCH codes. By Delsarte's theorem, we have (see Section A.2)

$$M(q)^{\perp} = \{(\mathrm{Tr}\,(ax + bx^{-1}))_{x \in \mathbf{F}_q^{\times}} : a, b \in \mathbf{F}_q\}.$$

This shows that $M(q)^{\perp}$ and therefore $M(q)$ do not depend on the choice of α up to equivalence. For a description of the weight distribution of $M(q)^{\perp}$, we need the Kronecker class number of binary quadratic forms.

Let Δ be a negative integer congruent to 0 or 1 (mod 4). We put

$$B(\Delta) = \{ax^2 + bxy + cy^2 \ : \ a, b, c \in \mathbf{Z}, a > 0, b^2 - 4ac = \Delta\}$$

and

$$b(\Delta) = \{ax^2 + bxy + cy^2 \in B(\Delta) \ : \ \gcd(a, b, c) = 1\}.$$

The members of $B(\Delta)$ are called *positive definite binary quadratic forms with discriminant* Δ. Those in $b(\Delta)$ are called *primitive*. The group $SL_2(\mathbf{Z})$ acts on $B(\Delta)$ by

$$\begin{pmatrix} \alpha & \beta \\ \gamma & \delta \end{pmatrix} \cdot f(x, y) = f(\alpha x + \beta y, \gamma x + \delta y)$$

for $f(x, y) \in B(\Delta)$. The subset $b(\Delta)$ is preserved under this action. Then the number of orbits in $b(\Delta)$ is called the *class number* of Δ and is denoted by $h(\Delta)$. The *Kronecker class number* of Δ is denoted by $H(\Delta)$; it is defined to be the number of orbits in $B(\Delta)$, but one should count the forms $ax^2 + ay^2$ and $ax^2 + axy + ay^2$, if at all present in $B(\Delta)$, with multiplicity 1/2 and 1/3 respectively. It is known that the relation between the Kronecker class numbers and the ordinary class numbers is given by the formula ([28])

$$H(\Delta) = \sum_f h_w \left(\frac{\Delta}{f^2}\right),$$

where f runs over all positive divisors of Δ for which $\Delta/f^2 \in \mathbf{Z}$ is congruent to 0 or 1 (mod 4) and h_w was defined in Theorem 3.4 by

$$h_w(-3) = \frac{1}{3}, \quad h_w(-4) = \frac{1}{2},$$
$$h_w(\Delta) = h(\Delta) \quad \text{if } \Delta < -4.$$

Next we consider a relation between certain codes and elliptic curves over finite fields. We denote by $C(q)$ the binary code of length $q - 1$ defined by

$$\{c(a, b) = (\mathrm{Tr}\,(ax + bx^{-1}))_{x \in \mathbf{F}_q^\times} \ : \ a \in \mathbf{F}_q, b \in \mathbf{F}_2\}.$$

Then we have

THEOREM 3.10 The weights of codewords $c(a, b)$ in $C(q)$ are given as follows:

(1) $c(0, 1)$ has weight $q/2$;

(2) $c(a, 0)$ has weight $q/2$ for $a \neq 0$;

(3) $c(a, 1)$ has weight $q - |E_a(\mathbf{F}_q)|/2$ for $a \neq 0$;

here E_a denotes the elliptic curve with the Weierstrass equation

$$E_a : y^2 + xy = x^3 + ax.$$

Proof. The weights of codewords $c(a, b)$ are easily determined whenever $ab = 0$. From Section 1.4 we know that the trace $\mathrm{Tr} : \mathbf{F}_q \to \mathbf{F}_2$ is a non-trivial linear form. Therefore the weight of $c(0, 1) = (\mathrm{Tr}\,(x^{-1}))_{x \in \mathbf{F}_q^\times}$ is the cardinality of the complement of the kernel of this linear form, hence equal to $q/2$. In the same way, for $a \neq 0$ we obtain $w(c(a, 0)) = w(c(1, 0)) = w(\mathrm{Tr}\,(x)_{x \in \mathbf{F}_q^\times}) = q/2$. In the remaining case the weight is closely related to the number of rational points on the curve

$$C_a : y^2 + y = ax + x^{-1}$$

over $\mathbf{F}_q$. For a given $x \in \mathbf{F}_q^\times$ the quadratic equation

$$y^2 + y = ax + x^{-1}$$

has two or no solutions in $\mathbf{F}_q$ depending on whether $\mathrm{Tr}\,(ax + x^{-1})$ equals 0 or 1. Let $\gamma = ax + x^{-1}$. If $y^2 + y = \gamma$ has a solution $y \in \mathbf{F}_q$, then $y + 1$ is the other solution, and property (6) in Section 1.4 yields $\mathrm{Tr}\,(\gamma) = \mathrm{Tr}\,(y^2 + y) = \mathrm{Tr}\,(y^2) + \mathrm{Tr}\,(y) = \mathrm{Tr}\,(y) + \mathrm{Tr}\,(y) = 0$. Now let $T_0 = \{\gamma \in \mathbf{F}_q : \mathrm{Tr}\,(\gamma) = 0\}$. Since the trace $\mathrm{Tr} : \mathbf{F}_q \to \mathbf{F}_2$ is onto, there are exactly $q/2$ elements in T_0. On the other hand, the map $y \mapsto y^2 + y$ from $\mathbf{F}_q$ to T_0 has the property that any $\gamma \in T_0$ has either two distinct preimages or no preimage at all in $\mathbf{F}_q$. But the latter case cannot occur because of $|\mathbf{F}_q| = 2|T_0|$. Therefore the map is onto, i.e., for every $\gamma \in T_0$ the equation $y^2 + y = \gamma$ has two solutions y in $\mathbf{F}_q$. Therefore

$$\mathrm{Tr}\,(ax + x^{-1}) = 1 - \frac{1}{2}|\{y \in \mathbf{F}_q : y^2 + y = ax + x^{-1}\}|,$$

and hence the weight of the codeword $c(a, 1)$ is equal to $q-1-|C_{\text{aff}}(\mathbf{F}_q)|/2$. Here $C_{\text{aff}}(\mathbf{F}_q)$ denotes the set of points on the curve C_a in the affine plane over $\mathbf{F}_q$. Multiplying the equation C_a by a^2x^2 and replacing y by $x^{-1}y$ and x by $a^{-1}x$, we find the equation of an elliptic curve

$$E_a : y^2 + xy = x^3 + ax.$$

Thus C_a and E_a have the same numbers of points in the projective plane over $\mathbf{F}_q$. Since C_a has two points at infinity, it follows that $|C_{\text{aff}}(\mathbf{F}_q)| = |E_a(\mathbf{F}_q)| - 2.$ ∎

To determine the weight distribution of the codes $C(q)$, we must count the number of isomorphism classes of elliptic curves over $\mathbf{F}_q$. Two elliptic curves are said to be *isomorphic* if they are isomorphic as projective varieties. The elliptic curve E is said to be *supersingular* if p divides t, where $|E(\mathbf{F}_q)| = q+1-t$. Otherwise, it is called *non-supersingular*. It is known that if $p = 2$ or $p = 3$, then E is non-supersingular if and only if $j(E) \neq 0$ ([6]). We denote by $M_q(t)$ the number of isomorphism classes of elliptic curves over $\mathbf{F}_q$ such that $|E(\mathbf{F}_q)| = q + 1 - t$. Then we have the following result.

PROPOSITION 3.11 Let $\mathbf{F}_q$ be an extension of $\mathbf{F}_2$ and let t be an odd integer. Then

$$M_q(t) = \begin{cases} H(t^2 - 4q) & \text{if } |t| < 2\sqrt{q}, \\ 0 & \text{otherwise.} \end{cases}$$

Proof. One can find a proof in [6]. ∎

As a consequence we find the following weight distribution for the code $C(q)$:

THEOREM 3.12 The non-zero weights in the code $C(q)$ are $w_t = (q - 1 + t)/2$, where $t \in \mathbf{Z}$, $t^2 < 4q$ and $t \equiv 1 \pmod 4$. Also the number of codewords in $C(q)$ of weight $(q - 1 + t)/2$ is given by

$$\begin{cases} H(t^2 - 4q) & \text{if } t \neq 1, \\ H(1 - 4q) + q & \text{if } t = 1. \end{cases}$$

Proof. Consider the elliptic curves

$$E_a : \ y^2 + xy = x^3 + ax \quad (a \in \mathbf{F}_q^\times).$$

We have $j(E_a) = a^{-2}$. Therefore every elliptic curve E_a is non-supersingular. Every element in $\mathbf{F}_q^\times$ occurs exactly once as a j-invariant in this family. It is known that over $\mathbf{F}_q$ there are, for every non-zero j-invariant, precisely two elliptic curves having this j-invariant, and if one of these has $q + 1 - t$ points over $\mathbf{F}_q$, then the other has $q + 1 + t$ points[1]. Since every curve in the family has $(a^{q/2}, 0)$ as an $\mathbf{F}_q$-rational point of order 4, we conclude that every elliptic curve E over $\mathbf{F}_q$ with 4 dividing $|E(\mathbf{F}_q)| = q + 1 - t$ occurs exactly once in our family. The number of elliptic curves over $\mathbf{F}_q$ with precisely $q + 1 - t$ points over $\mathbf{F}_q$ is equal to the Kronecker class number $H(t^2 - 4q)$ when $t^2 < 4q$ and t is odd. Therefore we have that the non-zero weight of $C(q)$ are the numbers

$$q - \frac{1}{2}(q + 1 - t) = \frac{1}{2}(q - 1 + t),$$

where $t^2 < 4q$ and $t \equiv 1 \pmod 4$. When $t \neq 1$, only codewords of type $c(a, 1)$ can have weight $(q - 1 + t)/2$, and there are precisely $H(t^2 - 4q)$ such codewords. When $t = 1$, the codewords $c(0, 1)$ and $c(a, 0)$ with $a \in \mathbf{F}_q^\times$ also have weight $q/2$; hence $H(1 - 4q) + q$ codewords have weight $q/2$. ∎

Now we apply Theorem 3.12 to the dual Melas code $M(q)^\perp$:

THEOREM 3.13 The non-zero weights of $M(q)^\perp$ are $w_t = (q - 1 + t)/2$, where $t \in \mathbf{Z}$, $t^2 < 4q$ and $t \equiv 1 \pmod 4$. For $t \neq 1$, the frequency of w_t is $(q-1)H(t^2-4q)$; the weight $w_1 = q/2$ has frequency $(q-1)(H(1-4q)+2)$.

Proof. The group $\mathbf{F}_q^\times$ acts on the code $M(q)^\perp$ by

$$(a, b) \mapsto (\zeta a, \zeta^{-1} b)$$

[1]R. Schoof: Nonsingular plane cubic curves over finite fields, J. Combin. Theory, A 46 (1987), 182-211.

for $\zeta \in \mathbf{F}_q^\times$. The codewords in the same orbit have the same weight. For $b = 0$ we find the zero codeword and $q - 1$ codewords of weight $q/2$ in $M(q)^\perp$. The set of codewords with $b \neq 0$ is stable under the action of $\mathbf{F}_q^\times$; every orbit has length $q - 1$ and contains exactly one codeword $c(a, 1)$ of the code $C(q)$. Therefore, apart from the weight $q/2$, the theorem follows from Theorem 3.12.

The $q - 1$ codewords with $b = 0$ and the $q - 1$ codewords in the orbit of $c(0, 1)$ all have weight $q/2$. Together with the $H(1 - 4q)$ orbits of codewords $c(a, 1)$ with $a \neq 0$ that have weight $q/2$ in $C(q)$, we find $(q - 1)H(1 - 4q) + 2(q - 1)$ codewords of weight $w_1 = q/2$. ∎

To obtain the weight distribution of the Melas codes themselves, we shall combine Theorem 3.13, the MacWilliams identities and the following Eichler-Selberg trace formula for $\Gamma_1(4)$. This level 4 is related to the fact that the curves in our family all admit an $\mathbf{F}_q$-rational point of order 4.

THEOREM 3.14 Let $m \geq 1$ and $q = 2^m$. The trace of the Hecke operator T_q acting on the space of cusp forms $S_k(\Gamma_1(4))$ of weight $k \geq 2$ is given by

$$
\mathrm{Tr}\,(T_q) = \begin{cases} -1 + q - \sum_t H(t^2 - 4q) = 0 & \text{if } k = 2, \\ -1 - (-1)^{kq/2} \sum_t Q_{k-2}(t, q) H(t^2 - 4q) & \text{if } k \geq 3, \end{cases}
$$

where t runs over $\{t \in \mathbf{Z} : t^2 < 4q, \; t \equiv q + 1 \pmod 4\}$ and the numbers $Q_k(t, n)$ are recursively defined by

$$Q_0(t, n) = 1, \; Q_1(t, n) = t,$$

$$Q_{k+1}(t, n) = t Q_k(t, n) - n Q_{k-1}(t, n)$$

for $k \geq 1$.

Proof. The result follows directly from the trace formula in Theorem 3.4. ∎

THEOREM 3.15 (SCHOOF AND VAN DER VLUGT [28], [29]) The number A_i of codewords of weight i in the Melas code $M(q)$ is given by

$$q^2 A_i = \binom{q-1}{i} + 2(-1)^{\lfloor \frac{i+1}{2} \rfloor}(q-1)\binom{q/2-1}{\lfloor i/2 \rfloor}$$

$$-(q-1) \sum_{\substack{j=0 \\ j \equiv i \pmod 2}}^{i} W_{i,j}(q)(1 + \tau_{j+2}(q)),$$

where $\lfloor x \rfloor$ is the greatest integer part ($\leq x$) of $x \in \mathbf{R}$, and $W_{i,j}(q)$ and $\tau_k(q)$ are defined as follows. The polynomials $W_{i,j}(q)$ are for $0 \leq j \leq i$ and $i \equiv j \pmod 2$ defined by

$$W_{0,0} = 1, \ W_{1,1} = 1,$$

$$(i+1)W_{i+1,j+1} = -qW_{i,j+2} - W_{i,j} - (q-i)W_{i-1,j+1},$$

and otherwise the $W_{i,j}$ are zero. If $k \geq 3$ then $\tau_k(q)$ denotes the trace of the Hecke operator T_q on the space $S_k(\Gamma_1(4))$, and we let $\tau_2(q) = -q$.

REMARK 3.6 ([21]) For any positive integer n, we define the *Krawtchouk polynomial*

$$P_k(x; n) = P_k(x) = \sum_{j=0}^{k}(-1)^j \binom{x}{j}\binom{n-x}{k-j},$$

where $k = 0, 1, \cdots, n$. These polynomials have the generating function

$$\sum_{k=0}^{\infty} P_k(x)z^k = (1+z)^{n-x}(1-z)^x.$$

Then the Krawtchouk polynomials satisfy the recurrence

$$(k+1)P_{k+1}(x) = (n-2x)P_k(x) - (n-k+1)P_{k-1}(x),$$

for $k = 1, 2, \cdots$, with $P_0(x) = 1$, $P_1(x) = n - 2x$. These polynomials will be used in the proof of Theorem 3.15.

REMARK 3.7 ([21]) For the polynomial $Q_k(t,n)$ defined in Theorem 3.14, when we assign the variable t a weight equal to 1 and n a weight equal to 2 then $Q_k(t,n)$ is seen to be homogeneous of weight k. Viewed as a polynomial in t it is monic, and therefore we can write

$$t^i = \sum_{\substack{j=0 \\ j:\text{even}}}^{i} \lambda_{i,j} Q_{i-j}(t,n) n^{j/2}, \qquad (3.3)$$

where the integers $\lambda_{i,j}$ satisfy

$$\lambda_{i,j} = 0 \text{ whenever } j \notin \{0,1,\cdots,i\} \text{ or } j \text{ odd},$$
$$\lambda_{0,0} = \lambda_{1,0} = 1,$$
$$\lambda_{i+1,j} = \lambda_{i,j-2} + \lambda_{i,j}.$$

We have also

$$\lambda_{i,2j} = \binom{i-1}{j} - \binom{i-1}{j-2}.$$

Proof of Theorem 3.15. For $0 \le l \le q-1$, we put

$$f_l(x) = P_l\left(\frac{q-1+x}{2}; q-1\right) = P_l\left(\frac{q-1+x}{2}\right).$$

From the recurrence relation of the Krawtchouk polynomial $P_l(x)$, we have

$$f_0(x) = 1, f_1(x) = -x,$$
$$(l+1)f_{l+1}(x) = -xf_l(x) - (q-1)f_{l-1}(x).$$

It follows that f_l has degree l and the parity of f_l is equal to the parity of l. Therefore we can write

$$f_l(x) = \sum_{\substack{k=0 \\ k \equiv l \ (\text{mod } 2)}}^{l} \pi_l(k) x^k, \qquad (3.4)$$

where $\pi_0(0) = 1$, $\pi_1(1) = -1$ and

$$(l+1)\pi_{l+1}(k+1) = -\pi_l(k) - (q-1)\pi_{l-1}(k+1).$$

Now we apply the MacWilliams identities to the weight distributions of the Melas code $M(q)$. First we have that

$$q^2 A_i = \sum_t \text{frequency}(w_t) P_i \left(\frac{q-1+t}{2} \right) + P_i(0),$$

where t runs over $\{t \in \mathbf{Z} : t^2 < 4q, t \equiv 1 \pmod 4\}$ ([21, Chap.5, formula (13)]). Using Theorem 3.13 and f_i above, we know

$$\frac{q^2}{q-1} A_i = \sum_t f_i(t) H(t^2 - 4q) + \frac{P_i(0)}{q-1} + 2f_i(1),$$

where

$$P_i(0) = \binom{q-1}{i}$$

and

$$f_i(1) = (-1)^{\lfloor (i+1)/2 \rfloor} \binom{\frac{q}{2}-1}{\lfloor i/2 \rfloor}.$$

From (3.4) we obtain that

$$\sum_t f_i(t) H(t^2 - 4q) = \sum_{\substack{j=0 \\ j \equiv i \ (\text{mod } 2)}}^{i} \pi_i(j) \sum_t t^j H(t^2 - 4q).$$

By (3.3) this becomes

$$\sum_{\substack{j=0 \\ j \equiv i \ (\text{mod } 2)}}^{i} \pi_i(j) \sum_{\substack{k=0 \\ k:\text{even}}}^{j} \lambda_{j,k} q^{\frac{k}{2}} \sum_t Q_{j-k}(t,q) H(t^2 - 4q).$$

By Theorem 3.14 this becomes, since 4 divides q,

$$\sum_{\substack{j=0 \\ j \equiv i \ (\text{mod } 2)}}^{i} \pi_i(j) \sum_{\substack{k=0 \\ k:\text{even}}}^{j} \lambda_{j,k} q^{\frac{k}{2}} (-1 - \tau_{j-k+2}(q))$$

which after changing the summation variables somewhat is seen to be equal to

$$\sum_{\substack{j=0 \\ j \equiv i \ (\text{mod } 2)}}^{i} W_{i,j}(q)(-1 - \tau_{j+2}(q)),$$

where

$$W_{i,j}(q) = \sum_{\substack{k=0 \\ k:\text{even}}}^{i-j} \pi_i(k+j)\lambda_{k+j,k}q^{k/2}$$

and $W_{i,j}(q)$ satisfy the recurrence relation as required. $\blacksquare$

REMARK 3.8 1) The main problem in applying Theorem 3.15 is the calculation of the $\tau_k(q)$ for $k \geq 3$ which is the trace of the Hecke operator T_q ($q = 2^m$, $m \geq 3$) acting on the space $S_k(\Gamma_1(4))$.

2) It would be interesting to extend the methods of Theorem 3.15 to other families of cyclic codes.

Chapter 4

ALGEBRAIC-GEOMETRIC CODES AND MODULAR CURVE CODES

1. Classical Goppa codes

The Russian mathematician Goppa invented an extended class of codes that contains the BCH codes as special codes. In this section we introduce these powerful and exciting classical Goppa codes.

1.1 The basic idea

Let C be a primitive BCH code in the narrow sense, with designed distance δ. It is defined by means of a primitive n-th root of unity β in some extension field of $\mathbf{F}_q$. Then a word $c = (c_0, c_1, \cdots, c_{n-1}) \in \mathbf{F}_q{}^n$ belongs to C if and only if

$$c_0 + c_1(\beta^j) + c_2(\beta^j)^2 + \cdots + c_{n-1}(\beta^j)^{n-1} = 0$$

for all j with $1 \leq j < \delta$. Hence

$$
\begin{aligned}
(x^n - 1) \sum_{i=0}^{n-1} \frac{c_i}{x - \beta^{-i}} &= \sum_{i=0}^{n-1} c_i \sum_{k=0}^{n-1} x^k (\beta^{-i})^{n-1-k} \\
&= \sum_{k=0}^{n-1} x^k \sum_{i=0}^{n-1} c_i (\beta^{k+1})^i \\
&= x^{\delta-1} p(x)
\end{aligned}
$$

49

for some polynomial $p(x)$. Therefore we have

$$\sum_{i=0}^{n-1} \frac{c_i}{x - \beta^{-i}} = \frac{x^{\delta-1}p(x)}{x^n - 1}.$$

We generalize this by replacing the polynomial $x^{\delta-1}$ by an arbitrarily chosen polynomial $g(x)$:

DEFINITION 4.1 Let $F \subseteq E$ be finite fields, let $g(x)$ be a polynomial over E, and let $L = \{\beta_0, \beta_1, \cdots, \beta_{n-1}\}$ be a set of elements of E such that $g(\beta_i) \neq 0$ for $0 \leq i \leq n - 1$. Then the *classical Goppa code* $\Gamma(L, g)$ is the set of words $(c_0, c_1, \cdots, c_{n-1})$ in F^n for which

$$\sum_{i=0}^{n-1} \frac{c_i}{x - \beta_i} \equiv 0 \ (\mathrm{mod} \ g(x)). \tag{4.1}$$

The polynomial $g(x)$ is called the *Goppa polynomial* of the code. If $E = F$, then we call $\Gamma(L, g)$ a *full Goppa code*, otherwise a *subfield Goppa code*.

RS codes are full Goppa codes with the Goppa polynomial $x^{\delta-1}$, and BCH codes are subfield Goppa codes with the same Goppa polynomial.

REMARK 4.1 Let $s(x)$ be a rational function, the representation $a(x)/b(x)$ of $s(x)$ is said to be *in lowest terms* if the highest common factor of the polynomials $a(x)$ and $b(x)$ is 1. If $g(x)$ is a polynomial then the congruence

$$s(x) \equiv 0 \ (\mathrm{mod} \ g(x))$$

means that in the representation of $s(x)$ as $a(x)/b(x)$ in lowest terms, $g(x)$ divides $a(x)$. We say that $g(x)$ divides the rational function $s(x)$ if $s(x) \equiv 0 \ (\mathrm{mod} \ g(x))$.

1.2 Basic properties of $\Gamma(L,g)$

Firstly we find a parity check matrix for $\Gamma(L,g)$. If $g(x)$ is any poly-nomial and $g(\beta) \neq 0$, we define $1/(x-\beta)$ to be the unique polynomial mod $g(x)$ such that

$$(x-\beta)\frac{1}{x-\beta} \equiv 1 \ (\mathrm{mod}\ g(x)),$$

namely

$$\frac{1}{x-\beta} = \frac{-1}{g(\beta)} \cdot \frac{g(x)-g(\beta)}{x-\beta}.$$

By (4.1) we see that

$$\left(\frac{1}{g(\beta_0)}\frac{g(x)-g(\beta_0)}{x-\beta_0}, \cdots, \frac{1}{g(\beta_{n-1})}\frac{g(x)-g(\beta_{n-1})}{x-\beta_{n-1}}\right)$$

with each entry interpreted as a column vector, is a parity check matrix for $\Gamma(L,g)$. Let

$$g(x) = \sum_{i=0}^{t} g_i x^i.$$

Then we have

$$\frac{g(x)-g(y)}{x-y} = \sum_{k+j \leq t-1} g_{k+j+1} y^j x^k,$$

so we can rewrite (4.1) as follows:

$$\sum_{i=0}^{n-1} c_i h_i \sum_{k+j \leq t-1} g_{k+j+1}(\beta_i)^j x^k = 0 \tag{4.2}$$

with $h_i = 1/g(\beta_i)$. In (4.2), the coefficient of x^k is 0 for $0 \leq k \leq t-1$. This means that if $c = (c_0, c_1, \cdots, c_{n-1})$ is a codeword, then it has inner product 0 with the rows of the matrix

$$\begin{pmatrix} h_0 g_t & h_1 g_t & \cdots & h_{n-1} g_t \\ h_0(g_{t-1}+g_t\beta_0) & h_1(g_{t-1}+g_t\beta_1) & \cdots & h_{n-1}(g_{t-1}+g_t\beta_{n-1}) \\ \vdots & & & \vdots \\ h_0(g_1+g_2\beta_0+\cdots+g_t\beta_0^{t-1}) & \cdots & \cdots & h_{n-1}(g_1+g_2\beta_{n-1}+\cdots+g_t\beta_{n-1}^{t-1}) \end{pmatrix}$$

Using elementary row operations this yields the following parity check matrix for $\Gamma(L,g)$:

$$H = \begin{pmatrix} h_0 & h_1 & \cdots & h_{n-1} \\ h_0\beta_0 & h_1\beta_1 & \cdots & h_{n-1}\beta_{n-1} \\ \vdots & \vdots & & \vdots \\ h_0\beta_0^{t-1} & h_1\beta_1^{t-1} & \cdots & h_{n-1}\beta_{n-1}^{t-1} \end{pmatrix}. \qquad (4.3)$$

We have also the following result.

THEOREM 4.1 Let $C = \Gamma(L,g)$ be a classical Goppa code with $|L| = n$ and $\deg g = t$. Then C is a linear code with length n, dimension $\geq n - mt$ and minimum distance $\geq t + 1$, where $E = \mathbf{F}_{q^m}$, $F = \mathbf{F}_q$.

Proof. The statement about the dimension follows from (4.3). If w is the weight of a codeword $c = (c_0, c_1, \cdots, c_{n-1})$, then the degree of the numerator $a(x)$ of the left hand side of (4.1) is $w - 1$. Since $g(x)$ divides $a(x)$, we get $w - 1 = \deg(a(x)) \geq \deg(g(x)) = t$. Therefore the minimum distance is at least $t + 1$. ∎

2.   Algebraic curves

The key of Goppa's definition of algebraic-geometric codes is to replace the set of polynomials over a finite field by a more general construction. He uses the language of algebraic curves to introduce the codes. In this section, we simply introduce some basic notions and properties of algebraic curves.

2.1   Affine varieties and projective varieties

Let k be an algebraically closed field. In our applications, k will be the algebraic closure of a finite field $\mathbf{F}_q$ with $q = p^r$ elements. The *affine n-space* over $\mathbf{F}_q$ is the set of n-tuples

$$\mathbf{A}^n = \mathbf{A}^n(k) = \{P = (x_1, \cdots, x_n) : x_i \in k\}.$$

The set of $\mathbf{F}_q$-rational points in $\mathbf{A}^n$ is

$$\mathbf{A}^n(\mathbf{F}_q) = \{P = (x_1, \cdots, x_n) \ : \ x_i \in \mathbf{F}_q\}.$$

In the space $\mathbf{A}^n$, an *affine algebraic set* is the set of zeros of an ideal I of $k[x_1, \cdots, x_n]$, namely

$$B = V(I) = \{P \in \mathbf{A}^n \ : \ f(P) = 0 \text{ for all } f \in I\}.$$

The ideal associated to an affine algebraic set V is the ideal

$$I(V) = \{f \in k[x_1, \cdots, x_n] \ : \ f(P) = 0 \text{ for all } P \in V\}.$$

An algebraic set B is called *irreducible* if B cannot be written as the union of two proper algebraic subsets of B. The set $V(I)$ is irreducible if and only if $I(V)$ is a prime ideal.

DEFINITION 4.2 An affine algebraic set V is called an *affine variety*, if $I(V)$ is a prime ideal in the ring $k[x_1, \cdots, x_n]$.

An affine algebraic set is defined over $\mathbf{F}_q$ if it can be defined by an ideal which is generated by elements in $\mathbf{F}_q[x_1, \cdots, x_n]$. An affine variety V is said to be *defined over* $\mathbf{F}_q$ if the underlying algebraic set is. If V is defined over $\mathbf{F}_q$ we call

$$\mathbf{F}_q[V] = \mathbf{F}_q[x_1, \cdots, x_n]/I(V/\mathbf{F}_q)$$

with $I(V/\mathbf{F}_q) = I(V) \cap \mathbf{F}_q[x_1, \cdots, x_n]$, the *affine coordinate ring* of V. The quotient field $\mathbf{F}_q(V)$ of $\mathbf{F}_q[V]$ is called the *function field* of V over $\mathbf{F}_q$.

For projective spaces the situation is more complicated because of the homogeneous coordinates. The *projective n-space* $\mathbf{P}^n$ over $\mathbf{F}_q$ is the set of all $(n+1)$-tuples

$$(x_0, x_1, \cdots, x_n) \in \mathbf{A}^{n+1} - \{0\}$$

modulo the following equivalence relation: we let $(x_0, x_1, \cdots, x_n) \sim (y_0, y_1, \cdots, y_n)$ if and only if $(x_0, x_1, \cdots, x_n) = \lambda(y_0, y_1, \cdots, y_n)$ for some $\lambda \in k^\times$. We write an element in $\mathbf{P}^n$ in homogeneous coordinates as

$(x_0 : x_1 : \cdots : x_n)$. A *projective algebraic set* is the set of zeros of a homogeneous ideal I of $k[x_0, x_1, \cdots, x_n]$,

$$V(I) = \{P \in \mathbf{P}^n \ : \ f(P) = 0 \text{ for all homogeneous } f \in I\}.$$

The *ideal* associated to a projective algebraic set V is the ideal

$$I(V) = \{f \in k[x_0, x_1, \cdots, x_n] \ : \ f(P) = 0 \text{ for all } P \in V\}.$$

DEFINITION 4.3 A projective algebraic set V is called a *projective variety*, if $I(V)$ is a homogeneous prime ideal.

A projective algebraic set V is *defined over* $\mathbf{F}_q$ if its defining ideal I is generated by polynomials in $\mathbf{F}_q[x_0, x_1, \cdots, x_n]$. If V is defined over $\mathbf{F}_q$, the $\mathbf{F}_q$-*rational points on* V consist of $V(\mathbf{F}_q) = V \cap \mathbf{P}^n(\mathbf{F}_q)$. Like in the affine case, a projective variety V is defined over $\mathbf{F}_q$, if this is the case for V seen as an algebraic set. Let V_1 and V_2 be projective varieties. A *rational map* from $V_1 \subseteq \mathbf{P}^m$ to $V_2 \subseteq \mathbf{P}^n$ is a map given by rational functions

$$\Phi_1 : V_1 \longrightarrow V_2$$

$$P = (x_0 : x_1 : \cdots : x_n) \mapsto (f_0(P) : f_1(P) : \cdots : f_n(P))$$

where $f_i \in k(V_1)$ are such that whenever they are all defined, the image $(f_0(P) : f_1(P) : \cdots : f_n(P))$ defines a point in V_2. If there exists a rational map

$$\Phi_2 : V_2 \longrightarrow V_1$$

such that $\Phi_2 \circ \Phi_1 = \Phi_1 \circ \Phi_2 = \text{id}$, Φ_1 (and Φ_2) are birational maps and we say that V_1 and V_2 are *birationally equivalent*. In case the compositions are defined everywhere, Φ_1 (and Φ_2) are isomorphisms and we say that V_1 and V_2 are *isomorphic*.

REMARK 4.2 We show how to embed an affine variety in a projective variety. Associated with $f \in k[x_1, \cdots, x_n]$ the homogeneous polynomial f^* is defined by

$$f^* = x_0^l f\left(\frac{x_1}{x_0}, \cdots, \frac{x_n}{x_0}\right),$$

where l is the degree of f. Let V be an affine variety in $\mathbf{A}^n$ defined by the prime ideal I. Let I^* be the ideal generated by $\{f^* : f \in I\}$. Then I^* is a homogeneous prime ideal defining the projective variety V^* in $\mathbf{P}^n$. We put $V_0^* = \{(x_0 : x_1 : \cdots : x_n) \in V^* : x_0 \neq 0\}$. Then V is isomorphic with V_0^* under the map

$$(x_1, \cdots, x_n) \mapsto (1 : x_1 : \cdots : x_n).$$

The points $(0 : x_1 : \cdots : x_n) \in V^*$ are called the points at infinity of V. Conversely, for any point P of a projective variety V and any hyperplane H not containing P, the complement $V - H$ is an affine variety containing P.

2.2 Divisors of algebraic curves

The dimension of the variety V is the transcendence degree of $k(V)$ over k. If this dimension is 1, V is called an *algebraic curve*. In the following we shall only consider algebraic curves C. For coding theory, one only wants to work with nice curves. The restriction we will need is that our curves will be nonsingular, a notion which we will define below.

Let $f(x, y) = \sum a_{ij} x^i y^j \in k[x, y]$. Then f_x, the *partial derivative* of f with respect to x, is defined by

$$f_x(x, y) = \sum i a_{ij} x^{i-1} y^j,$$

and f_y is defined similarly. Consider an algebraic curve C in $\mathbf{A}^2$ defined by an equation $f(x, y) = 0$, and let $P = (a, b)$ be a point on the curve C. If at least one of the derivatives f_x or f_y is not zero in P, then P is called a *simple* or *nonsingular* point of the curve C. A curve is called *nonsingular* or *smooth* if all the points are nonsingular. In this case, the curve has a tangent at P with equation

$$f_x(P)(x - a) + f_y(P)(y - b) = 0$$

for all of its points $P = (a, b)$. The definitions for a projective plane curve are similar.

From now on, we assume all curves to be smooth and projective.

Let C be a curve defined over $\mathbf{F}_q$. A *divisor* on C is a formal sum $D = \sum_{P \in C} n_P P$ with $n_P \in \mathbf{Z}$ and $n_P = 0$ for all but a finite number of points P. The *support of a divisor* is the set of points with nonzero coefficient. A divisor D is called *effective* if all coefficients n_P are nonnegative, and we denote this by $D \geq 0$. The *degree* $\deg(D)$ of the divisor D is $\sum n_P \deg(P)$. A *point of degree* n on C over $\mathbf{F}_q$ is a set $P = \{P_0, \cdots, P_{n-1}\}$ of n distinct points in $C(\mathbf{F}_{q^n})$ such that $P_i = \sigma_{q,n}^i(P_0)$ for $i = 1, \cdots, n-1$, where $\sigma_{q,n}$ denotes the Frobenius automorphism on $\mathbf{F}_{q^n}$ defined by $\sigma_{q,n}(\alpha) = \alpha^q$. Let $\mathrm{Div}\,(C)$ be the set of divisors on C. We have a *partial ordering* $\leq$ on $\mathrm{Div}\,(C)$ defined by

$$D \leq D' \text{ if and only if } n_P \leq n_P' \text{ for all } P \in C,$$

where $D' = \sum n_P' P$. Furthermore, $\mathrm{Div}\,(D)$ carries a natural structure as an abelian group.

Next we define *the divisor of a rational function* f on C. If C and C' are two distinct projective plane curves over $\mathbf{F}_q$ defined by polynomials of degree d and e respectively, then points over $k = \overline{\mathbf{F}_q}$ in which they intersect will be denoted as a sequence of points $P_1, \cdots, P_l$ of varying degrees over $\mathbf{F}_q$, where a point is listed repeatedly according to the multiplicity of intersection of the two curves at this point. (The concept of multiplicity of intersection will not be explained here.) Further, we have $de = r_1 + \cdots + r_l$, where r_i is the degree of P_i over $\mathbf{F}_q$. To express this, we might write

$$C \cap C' = P_1 + \cdots + P_l$$

and call $C \cap C'$ the *intersection divisor* of C and C'. Note that the intersection divisor $C \cap C'$ is an effective divisor of degree de. Let $\varphi(x, y, z)$ be the polynomial which defines the curve C over $\mathbf{F}_q$. Then the function field of C over $\mathbf{F}_q$ is

$$\mathbf{F}_q(C) =$$

$$\left(\left\{ \frac{g(x,y,z)}{h(x,y,z)} : \begin{array}{l} g, h \in \mathbf{F}_q[x,y,z] \text{ are homogeneous} \\ \text{of the same degree} \end{array} \right\} \cup \{0\} \right) / \sim,$$

where $g/h \sim g'/h'$ if and only if $gh' - g'h \in (\varphi) \subseteq \mathbf{F}_q[x,y,z]$. The elements of $\mathbf{F}_q(C)$ are called *rational functions* on C. Let $f = g/h \in \mathbf{F}_q(C)$. Then we have that the curve C_g defined by $g = 0$ and the curve

C_h defined by $h = 0$ each intersect C in exactly de points of $\mathbf{P}^2(k)$, where d is the degree of the polynomial defining C and $e = \deg g = \deg h$ (by Bezout's theorem). Then the *divisor* of f is defined to be

$$\operatorname{div}(f) = \sum P - \sum Q,$$

where $\sum P$ is the intersection divisor $C \cap C_g$ and $\sum Q$ is the intersection divisor $C \cap C_h$.

The points where C and C_g intersect are the zeros of f, and the points where C and C_h intersect are the poles of f. So we think of $\operatorname{div}(f)$ as being the zeros of f minus the poles of f. Since $\deg(C \cap C_g) = \deg(C \cap C_h) = de$, we have $\deg \operatorname{div}(f) = 0$. Therefore f has the same number of zeros as poles. The divisor $\operatorname{div}(f)$ is called a *principal divisor*. We shall call two divisors D and D' *linearly equivalent* if $D - D'$ is a principal divisor and denote it by $D \equiv D'$. This is indeed an equivalence relation.

Now that we know what divisors, rational functions and divisors of rational functions are, we are ready for the following definition.

DEFINITION 4.4 Let D be a divisor on the curve C defined over $\mathbf{F}_q$. Then we define the subspace $L(D)$ of rational functions associated to D by

$$L(D) = \{f \in \mathbf{F}_q(C)^\times : \operatorname{div}(f) + D \geq 0\} \cup \{0\},$$

and we put

$$l(D) = \dim L(D).$$

Note that if $D \equiv D'$ and g is a rational function with $\operatorname{div}(g) = D - D'$, then the map $f \mapsto fg$ shows that $L(D)$ and $L(D')$ are isomorphic.

Calculating the value of $l(D)$ is quite difficult. It does not behave as regularly as one might expect. The following Riemann theorem, which is the most important result in algebraic geometry, gives a general estimate for $l(D)$ in terms of a constant g, called the *genus of the curve*. Using this theorem greatly simplifies the calculations required to determine $L(D)$. And it is also the key to the new results in coding theory.

THEOREM 4.2 (RIEMANN) Let C be a nonsingular projective plane curve defined over the finite field $\mathbf{F}_q$. Then there exists uniquely a non-negative number g such that for all divisors D with $\deg(D) > 2g - 2$,

$$l(D) = \deg(D) + 1 - g.$$

We do not give the proof; see [10] for a proof.

3.   The zeta functions of curves and rational points

3.1   Basic properties of the zeta functions

In this section, we also assume all curves to be smooth and projective. Let C be a curve defined over $\mathbf{F}_q$ and let N_m denote the number of $\mathbf{F}_{q^m}$-rational points on C. The *zeta function* of C over $\mathbf{F}_q$ is the formal power series

$$Z(C, t) = \exp\left(\sum_{m=1}^{\infty} N_m \frac{t^m}{m}\right).$$

We see that $Z(C, t)$ stores information on the numbers N_m for all $m \geq 1$. The case of C being an elliptic curve was treated in Chapter 3 (Remark 3.3).

REMARK 4.3 The Galois group $G = \mathrm{Gal}(k/\mathbf{F}_q)$ acts on $\mathbf{P}^n$ by acting on the homogeneous coordinates

$$\sigma \cdot (x_0 : x_1 : \cdots : x_n) = (\sigma(x_0) : \sigma(x_1) : \cdots : \sigma(x_n))$$

for $\sigma \in G$. Hence

$$\mathbf{P}^n(\mathbf{F}_q) = \{P \in \mathbf{P}^n \ : \ \sigma \cdot P = P \text{ for all } \sigma \in G\}.$$

The number of $\mathbf{F}_q$-rational points in $\mathbf{P}^n$ is $(q^{n+1} - 1)/(q - 1)$. As $C \subseteq \mathbf{P}^2$, N_m is less than or equal to the number of $\mathbf{F}_{q^m}$-rational points in $\mathbf{P}^2$, that is

$$N_m \leq q^{2m} + q^m + 1 < 3q^{2m}.$$

So for $m \geq 3$, we have $N_m/m < q^{2m}$, and the series

$$\sum_{m=1}^{\infty} N_m \frac{t^m}{m}$$

has radius of convergence at least q^{-2}, which makes $Z(C,t)$ an analytic function on the open disc with this radius. We also have

$$\frac{d}{dt}\left\{\log(Z(C,t))\right\} = \frac{Z(C,t)'}{Z(C,t)} = \sum_{m=1}^{\infty} N_m t^{m-1}. \tag{4.4}$$

The following theorem describes some very strong properties of $Z(C,t)$:

THEOREM 4.3 (WEIL) Let C be a curve of genus g defined over $\mathbf{F}_q$, and let $Z(C,t)$ be its associated zeta function. Then we have:

(1) The zeta function may be written as

$$Z(C,t) = \frac{P(t)}{(1-t)(1-qt)},$$

where $P(t) \in \mathbf{Z}[t]$ is of degree $2g$.

(2) The zeta function satisfies the functional equation

$$Z\left(C, \frac{1}{qt}\right) = q^{1-g}t^{2-2g} Z(C,t).$$

(3) The polynomial $P(t)$ may be factored as

$$P(t) = \prod_{i=1}^{g}(1 - \alpha_i t)(1 - \overline{\alpha}_i t)$$

with algebraic integers α_i satisfying $|\alpha_i|^2 = q$ for $i = 1, \cdots, g$.

For the proof, see [9].

COROLLARY 4.4 With the notations as above, we have

$$N_m = 1 + q^m - \sum_{i=1}^{g}(\alpha_i^m + \overline{\alpha}_i^m).$$

Proof. By taking the logarithmic derivative of

$$Z(C,t) = \frac{\prod_{i=1}^{2g}(1 - \alpha_i t)}{(1 - t)(1 - qt)},$$

(4.4) gives

$$
\begin{aligned}
\sum_{m=1}^{\infty} N_m t^{m-1} &= \frac{1}{1-t} + \frac{q}{1-qt} + \sum_{i=1}^{2g} \frac{-\alpha_i}{1 - \alpha_i t} \\
&= \sum_{m=0}^{\infty} t^m + \sum_{m=0}^{\infty} q^{m+1} t^m + \sum_{i=1}^{2g} \left(-\sum_{m=0}^{\infty} \alpha_i^{m+1} t^m \right).
\end{aligned}
$$

Now compare coefficients. ∎

COROLLARY 4.5 Let C be a curve of genus g defined over $\mathbf{F}_q$. Then the number N_m is bounded by

$$|N_m - 1 - q^m| \le 2g q^{m/2}.$$

Proof. From Corollary 4.4, we have

$$
\begin{aligned}
|N_m - 1 - q^m| &= \left| \sum_{i=1}^{2g} \alpha_i^m \right| \\
&\le \sum_{i=1}^{2g} |\alpha_i|^m \\
&= 2g q^{m/2},
\end{aligned}
$$

the last equality coming from (3) of Theorem 4.3. ∎

REMARK 4.4 The zeta function $Z(C,t)$ is holomorphic in the complex plane except in $t = 1$ and $t = 1/q$ where it has simple poles, and the zeros of $Z(C,t)$ are denoted by $\alpha_1^{-1}, \cdots, a_{2g}^{-1}$.

EXAMPLE 4.1 Consider the elliptic curve C in $\mathbf{P}^2$ defined over $\mathbf{F}_2$ by the equation

$$x_0^3 + x_1^3 + x_2^3 = 0.$$

The curve has genus $g = 1$. The set of $\mathbf{F}_2$-rational points is

$$\{(0:1:1),(1:1:0),(1:0:1)\},$$

which gives $N_1 = 3$. Therefore, with a defined as in Remark 3.3,

$$3 = N_1 = \frac{d}{dt}\{\log(Z(C,t))\}|_{t=0}$$

$$= \left[\frac{a+4t}{1+at+2t^2} + \frac{1}{1-t} + \frac{2}{1-2t}\right]_{t=0}.$$

Hence $a = 0$ and

$$P(t) = 1 + 2t^2 = (1 - i\sqrt{2}t)(1 + i\sqrt{2}t).$$

Now Corollary 4.4 implies

$$N_m = 1 + 2^m - (i\sqrt{2})^m - (-i\sqrt{2})^m,$$

and thereby we have

$$N_m = \begin{cases} 1 + 2^m, & \text{if } m \equiv 1 \pmod 2, \\ 1 + 2^m + 2(\sqrt{2})^m, & \text{if } m \equiv 2 \pmod 4, \\ 1 + 2^m - 2(\sqrt{2})^m, & \text{if } m \equiv 0 \pmod 4. \end{cases}$$

3.2 Maximum number of rational points

In 1980, Goppa came up with the beautiful idea of an error-correcting code to a linear system on a curve over a finite field. In order to construct good codes one needs curves with many rational points, and Goppa's work led to a revival of interest in rational points on curves over finite fields. Let C be a smooth projective curve of genus g over $\mathbf{F}_q$. It is also assumed that the curve C is *absolutely irreducible*; this means that the defining ideal is also prime in $k[x, y, z]$. We denote by $N_q(g)$ the *maximum number of rational points* on a curve of genus g over $\mathbf{F}_q$, namely

$$N_q(g) = \max\{|C(\mathbf{F}_q)| \; : \; C \text{ is a curve over } \mathbf{F}_q \text{ of genus } g\}.$$

Then the *Hasse-Weil bound* (Corollary 4.5) implies

$$N_q(g) \leq q + 1 + \lfloor 2g\sqrt{q}\rfloor, \tag{4.5}$$

where $\lfloor x \rfloor$ is the integer part $(\leq x)$ of $x \in \mathbf{R}$. Let

$$A(q) = \limsup_{g \to \infty} \frac{N_q(g)}{g}.$$

It follows from (4.5) that

$$A(q) \leq 2\sqrt{q}.$$

This has been improved to the *Drinfeld-Vladut bound*:

$$A(q) \leq \sqrt{q} - 1. \tag{4.6}$$

Furthermore equality holds if $q = p^r$ is an even power of prime p. The equality is proved by studying the number of rational points of modular curves over finite fields ([24], [32]).

4.   Algebraic-geometric codes

4.1   Algebraic-geometric codes of the first kind

To avoid confusion, the letter C will be reserved in this section to refer to codes, while the letter X will be used for curves. Also, we will always be working over the finite field $\mathbf{F}_q$, so the symbol k can unambiguously be used to denote a positive integer (the dimension of a code) as in the earlier chapters on coding theory.

Let X be an absolutely irreducible smooth projective curve of genus g over $\mathbf{F}_q$. Assume that $P_1, \cdots, P_n$ are $\mathbf{F}_q$-rational points on the curve X and let $D = P_1 + \cdots + P_n$. Assume that G is a divisor on X with support consisting of only $\mathbf{F}_q$-rational points and disjoint from D, i.e., G contains P_i for $i = 1, \cdots, n$ with coefficients zero. Then no P_i can be a pole of any $f \in L(G)$, and $f(P_i) \in \mathbf{F}_q$ for any $f \in L(G)$ and any P_i.

DEFINITION 4.5 The linear code $C(D, G)$ over $\mathbf{F}_q$ is the image of the linear map

$$\alpha : L(G) \longrightarrow \mathbf{F}_q{}^n$$

defined by

$$\alpha(f) = (f(P_1), \cdots, f(P_n)).$$

The codes obtained in this way are called *algebraic-geometric codes of the first kind*. The following theorem gives the parameters of the code $C(D, G)$.

THEOREM 4.6 Suppose that $2g - 2 < \deg(G) < n$. Then the code $C(D, G)$ has parameters $[n, k, d]$ with

$$
\begin{aligned}
n &= \deg(D), \\
k &= \deg(G) - g + 1 \\
d &\geq \delta_1 = n - \deg(G).
\end{aligned}
$$

Proof. Since $L(G)$ is a linear space over $\mathbf{F}_q$ and the map α is a linear transformation, we see that $C = C(D, G)$ is a linear code. Further, its length is obviously $n = \deg(D)$. Clearly, the dimension k of the code is at most $\dim L(G)$. It is exactly $\dim L(G)$ if and only if α is one-to-one. This is true if and only if the kernel of α is trivial. So suppose $\alpha(f) = 0$. Then $f(P_1) = \cdots = f(P_n) = 0$, so the coefficient of each P_i in the divisor $\operatorname{div}(f)$ is at least 1. Since no P_i is in the support of G, we have that $\operatorname{div}(f) + G - D \geq 0$, which means that $f \in L(G - D)$. If we add the hypothesis that $\deg(G) < n$, then the divisor $G - D$ has negative degree, so its associated space of rational functions is $\{0\}$. This means $f = 0$, so $k = \dim C = \dim L(G)$. That $\dim L(G) = \deg(G) + 1 - g$ is exactly the statement of the Riemann theorem (Theorem 4.2), since $\deg(G) > 2g - 2$.

To get the lower bound on the minimum distance d of C, let $\alpha(f) = (f(P_1), \cdots, f(P_n)) \in C$ be a codeword of minimum non-zero weight d, that is, $\alpha(f)$ is zero at $n - d$ positions. This implies that f vanishes at certain points

$$
P_{i_1}, \cdots, P_{i_{n-d}},
$$

and thus the divisor

$$
\operatorname{div}(f) + G - P_{i_1} - \cdots - P_{i_{n-d}}
$$

is effective. Since $\deg(\operatorname{div}(f)) = 0$, it follows that $\deg(G) - (n - d) \geq 0$. In other words, we have

$$
d \geq n - \deg(G) = \delta_1. \quad \blacksquare
$$

The parameter δ_1 is called *the designed distance of C*. For a given field size q and genus g, the number of rational points on a projective curve is bounded above by the Hasse-Weil bound. (Corollary 4.5). It is clear that for a given field size q, genus g and information rate $R = k/n$, the larger the n, the larger the designed distance δ_1 of $C(D, G)$ will be. Thus in designing algebraic geometric codes, we are interested in curves that have the maximum possible number of rational points. By Theorem 4.6, we have also

$$R + \delta \geq 1 + \frac{1-g}{n},$$

where $\delta = d/n$ is called the *relative minimum distance*. Let $C = C(D, G)$ be an algebraic geometric code and let $f_1, \cdots, f_k$ be a basis for the linear space $L(G)$ over $\mathbf{F}_q$. Under the conditions of the theorem, we know that $\dim C = k$, and so we know that $\alpha(f_1), \cdots, \alpha(f_k)$ is a basis for C. This means that the matrix

$$\begin{pmatrix} f_1(P_1) & \cdots\cdots & f_1(P_n) \\ f_2(P_1) & \cdots\cdots & f_2(P_n) \\ \vdots & & \vdots \\ f_k(P_1) & \cdots\cdots & f_k(P_n) \end{pmatrix}$$

is a generator matrix for C.

The linear codes obtained from algebraic curves include many well-known codes such as RS codes, BCH codes and classical Goppa codes ([25]). The following example shows how RS codes are obtained from the projective line.

EXAMPLE 4.2 Goppa's construction may be seen as a generalization of the construction of RS codes. The simplest curve X we could use in Goppa's construction is the projective line $\mathbf{P}^1$ over $\mathbf{F}_q$, for which the genus is $g = 0$. $\mathbf{P}^1$ has exactly $q + 1$ points rational over $\mathbf{F}_q$. In homogeneous coordinates they are $(x : 1)$ where $x \in \mathbf{F}_q$, and the point at infinity, $Q = (1 : 0)$. We write $P_i = (\alpha^i : 1)$ for $i = 0, 1, \cdots, q-2$, where α is a primitive element of $\mathbf{F}_q^\times$. For the divisor D in Goppa's construction, take $\sum_{i=0}^{q-2} P_i$, and take $G = (k-1)Q$ with some integer $k < q - 1$. A basis for the linear space $L(G)$ in this case is $\{1, x, x^2, \cdots, x^{k-1}\}$. So $C(D, G)$ is an RS code.

EXAMPLE 4.3 Let X be an elliptic curve over the field $\mathbf{F}_q$. Assume that $P_1, \cdots, P_n$ are $\mathbf{F}_q$-rational points on X and let $D = P_1 + \cdots + P_n$. In order to pick a specific code we have to choose a divisor G such that G has support disjoint from D. One possible choice for the divisor G is $G = mQ$, where $Q = (0 : 1 : 0)$ is the point at infinity on X and $0 < m < n$. Since X has genus $g = 1$, the code $C(D, G)$ has parameters $[n, m, d]$ with $d \geq n - m$.

4.2 Algebraic-geometric codes of the second kind

Next we define another class of codes obtained from the curve X. We use the same notations as above.

DEFINITION 4.6 The linear code $C^*(D, G)$ over $\mathbf{F}_q$ is the set of vectors $(c_1, \cdots, c_n)$ such that $\sum_{i=1}^{n} c_i f(P_i) = 0$ for all $f \in L(D)$, namely

$$C^*(D, G) = \{(c_1 \cdots, c_n) \in \mathbf{F}_q{}^n : \sum_{i=1}^{n} c_i f(P_i) = 0 \text{ for all } f \in L(D)\}.$$

The codes obtained in this way are called *algebraic-geometric codes of the second kind*. It is easy to see from the definition that the codes $C(D, G)$ and $C^*(D, G)$ are dual to each other. If G is a non-negative multiple of a single rational point we shall call the code a *one-point code*. The following theorem gives the parameters of the code $C^*(D, G)$:

THEOREM 4.7 Assume that $2g - 2 < \deg(G)$. Then for the length n, dimension k' and minimum distance d' of $C^*(D, G)$, we have

$$\begin{aligned} n &= \deg(D), \\ k' &= n - \deg(G) + g - 1 + \dim L(G - D), \\ d' &\geq \delta_2 = \deg(G) + 2 - 2g. \end{aligned}$$

Proof. We prove only the last assertion for d'. Suppose that $c = (c_1, \cdots, c_n)$ is a non-zero codeword of C^* and arrange the places P_j so

that $c_j \neq 0$ for $j = 1, \cdots, w$ and $c_j = 0$ for $j > w$. Put $D_j = \sum_{i=1}^{j} P_i$.
We shall show that the assumption that $1 \leq w < \deg(G) - (2g - 2)$ leads
to a contradiction. For in that case, we have $\deg(G - D_w) > 2g - 2$, and
also $\deg(G - D_{w-1}) > 2g - 2$. Hence by Riemann's theorem (Theorem
4.2),

$$\dim L(G - D_w) = \deg(G) - w + 1 - g$$

and

$$\dim L(G - D_{w-1}) = \deg(G) - w + 2 - g.$$

Thus there exists some $f \in L(G - D_{w-1})$ for which $f \notin L(G - D_w)$.
This implies that $f(P_j) = 0$ for $j = 1, \cdots, w - 1$ and $f(P_w) \neq 0$. As
$G - D_{w-1} \leq G$, $f \in L(G)$ and

$$\sum_{j=1}^{n} c_j f(P_j) = c_w f(P_w) \neq 0,$$

contradicting the assumption that $c \in C^*$. ∎

The parameter δ_2 is called the *designed distance of* C^*. We will reveal
the fact that the distinction between C and C^* is spurious. In fact, there
exists a divisor K of degree $2g - 2$ such that $C^*(D, G) = C(D, K + D - G)$.

5.   Modular curves and codes

Our primary goal in this section is to make an outline of the main
applications of modular curves to the construction of algebraic-geometric
codes.

THEOREM 4.8 (TSFASMAN, VLADUT AND ZINK [35]) Let $\tilde{X}_0(N)$ be a
modular curve defined over $\mathbf{F}_q$ of genus $g_0(N)$. Then, for $q = p^2$ we have

$$\lim_{N \to \infty} \frac{g_0(N)}{|\tilde{X}_0(N)(\mathbf{F}_q)|} = \frac{1}{\sqrt{q} - 1}, \tag{4.7}$$

where N runs over a set of primes different from p.

For the proof of this theorem, the following two results are essential

REMARK 4.5 Let $q = p^{2m}$ be fixed, let $N(X)$ be the number of $\mathbf{F}_q$-rational points on a curve X over $\mathbf{F}_q$ and let $g(X)$ be its genus. Then we have

$$\liminf \frac{g(X)}{N(X)} = \frac{1}{\sqrt{q}-1},$$

where the lim inf is taken over all absolutely irreducible smooth projective algebraic curves over $\mathbf{F}_q$.

5.1 Modular curves and their reduction modulo p

We use the notation $\mathfrak{H}^* = \mathfrak{H} \cup \mathbf{Q} \cup \{\infty\}$ from Subsection 3.2.2. It is well-known that the orbit space $X_0(N) = \mathfrak{H}^*/\Gamma_0(N)$ carries the structure of a compact Riemann surface in a natural way. We will explain that $X_0(N)$ is also an algebraic curve defined over $\mathbf{Q}$. It is called a modular curve. The most important property of the modular curve $X_0(N)$ is the fact that $X_0(N)$ has a nonsingular projective model which is defined by an equation over $\mathbf{Q}$ whose reductions modulo primes p with $p \nmid N$, are also nonsingular (Igusa). In the following we will give a sketch of these ideas.

Let N be a positive integer and let $E = \mathbf{C}/L$ be an elliptic curve with period lattice $L = \mathbf{Z}\omega_1 + \mathbf{Z}\omega_2$ with $\omega_2/\omega_1 \in \mathfrak{H}$. Let C_N be a cyclic subgroup of E of order N which we may identify with the integral multiples of $\omega_2/N \bmod L$. Then we have the isomorphism

$$f : (E, C_N) \longrightarrow (E(z), C_N(z))$$

such that $f(C_N) = C_N(z)$, where $E(z) = \mathbf{C}/L_z$ with $L_z = \mathbf{Z}z + \mathbf{Z}$ $(z \in \mathfrak{H})$ and $C_N(z) = (\mathbf{Z}z + (1/N)\mathbf{Z})/L_z$. Also we know that two pairs $(E(z), C_N(z))$ and $(E(z'), C_N(z'))$ are isomorphic if and only if there exists an element $s \in \Gamma_0(N)$ such that $s(z) = z'$. Therefore we have the bijection

$$\mathfrak{H}^*/\Gamma_0(N) \longrightarrow \left\{\begin{array}{l} \text{elliptic} \quad \text{curves} \\ \text{over } \mathbf{C} \quad \text{and} \quad \text{a} \\ \text{cyclic} \quad \text{subgroup} \\ C_N \text{ in } E \text{ of order} \\ N \end{array}\right\} \Big/ \quad \text{modulo isomorphism}$$

$$\begin{array}{ccc} \cup & & \cup \\ z & \longmapsto & (E(z), C_N(z)) \end{array}$$

This bijection gives the interpretation of $X_0(N)$ as a moduli space of elliptic curves. Now it is known that the field of modular functions on $\Gamma_0(N)$ over $\mathbf{Q}$ is generated by $j(z)$ and $j(Nz)$, where $j(z)$ denotes the j-invariant function of $E(z)$. Therefore, the defining equation of the curve $X_0(N)$ is given by the modular equation

$$\Phi(j(z), j(Nz)) = 0.$$

By Igusa's theorem[1], we know that the modular curve $X_0(N)$ has a nonsingular projective model which is defined by an equation over $\mathbf{Q}$ whose reductions modulo primes p with $p \nmid N$ are also nonsingular. This nonsingular modular curve $\tilde{X}_0(N)$ defined over $\mathbf{F}_p$ is called a *reduction of $X_0(N)$ modulo p*.

REMARK 4.6 Put

$$A = \left\{ \begin{pmatrix} a & b \\ 0 & d \end{pmatrix} : ad = N, \ 0 \le b < d, \ (a, b, d) = 1, \ a, b, d \in \mathbf{Z} \right\},$$

and consider the polynomial

$$\Phi_N(t) = \prod_{\alpha \in A} (t - j(\alpha(z))).$$

We may view $\Phi_N(t)$ as a polynomial in two independent variables t and j over $\mathbf{Z}$, and write it as

$$\Phi(t, j) = \Phi_N(t, j) \in \mathbf{Z}[t, j].$$

Then the equation $\Phi(t, j) = 0$ is called the *modular equation of order N*.

---

[1] J. Igusa : Kroneckerian model of fields of elliptic modular functions, Amer. J. Math., 81 (1959), 561–577.

5.2 Proof of Theorem 4.8

We defined the Hecke operators in Section 3.2.4 which satisfy the relations

$$T_{p^r} = T_p \cdot T_{p^{r-1}} - p T_{p^{r-2}}, \quad T_1 = 1 \qquad (4.8)$$

for the powers of all primes $p \nmid N$.

It is well known that the cusp forms of weight 2 on $\Gamma_0(N)$ form a linear space $S_2(\Gamma_0(N))$ of dimension $g_0(N)$, where $g_0 = g_0(N)$ is the genus of $X_0(N)$. On the space $S_2(\Gamma_0(N))$, there is a *positive definite inner product* (the *Petersson inner product*)defined by

$$\langle f, g \rangle = \int_{\mathfrak{H}/\Gamma_0(N)} f(z)\overline{g(z)}dxdy \quad (x = \operatorname{Re}(z), y = \operatorname{Im}(z)),$$

where $\mathfrak{H}/\Gamma_0(N)$ denotes a *fundamental domain* of $\Gamma_0(N)$. The Hecke operators T_m for $\gcd(m, N) = 1$ are hermitian with respect to this inner product. Therefore, we can find a basis of common eigenforms for all T_m with $\gcd(m, N) = 1$, say

$$\{f_1, \cdots, f_{g_0}\}.$$

We can write

$$f_j(z) = \sum_{n=1}^{\infty} b_j(n)e^{2\pi i n z} = \sum_{n=1}^{\infty} b_j(n)q^n$$

with $b_j(1) = 1$, and then we have

$$f_j|_2 T_m = b_j(m)f_j \quad (1 \le j \le g_0)$$

for all m with $\gcd(m, N) = 1$. The basic fact is that the zeta function of $\tilde{X}_0(N)$ over $\mathbf{F}_p$ is expressible in terms of the Fourier coefficients $b_i(n)$ (*Eichler-Shimura congruence relation*): Let p be a prime with $(p, N) = 1$. Then the zeta function $Z(\tilde{X}_0(N), t)$ of $\tilde{X}_0(N)$ over $\mathbf{F}_p$ is of the form

$$Z(\tilde{X}_0(N), t) = \frac{\prod_{j=1}^{g_0}(1 - b_j(p)t + pt^2)}{(1 - t)(1 - pt)}.$$

By Corollary 4.4, we have

$$\alpha_j + \overline{\alpha}_j = b_j(p)$$

for $1 \leq j \leq g_0$. Therefore, if N is prime,

$$\left|\tilde{X}_0(N)(\mathbf{F}_p)\right| = p + 1 - \sum_{j=1}^{g_0} b_j(p)$$

and

$$\left|\tilde{X}_0(N)(\mathbf{F}_{p^2})\right| = p^2 + 1 - \sum_{j=1}^{g_0} b_j^2(p) + 2pg_0.$$

By (4.8), we have $b_j(p^2) = b_j^2(p) - p$. Hence

$$\left|\tilde{X}_0(N)(\mathbf{F}_{p^2})\right| = p^2 + 1 - \sum_{j=1}^{g_0} b_j(p^2) + pg_0.$$

On the other hand, the trace of the Hecke operator T_m was given in Theorem 3.4 in Section 3.2.4. Also we have that

$$\left\lfloor \frac{N}{12} \right\rfloor - 1 \leq g_0 \leq \left\lfloor \frac{N}{12} \right\rfloor + 1$$

by the *Hurwitz-Zenthen formula*. Now in the Eichler-Selberg trace formula (Theorem 3.4), we take N as a prime number different from p, $k = 2$, $m = p^2$ and $\chi = 1$. Then we have $\text{Tr}\,(T_{p^2}) = p^2 + g_0(N) - \{\text{sum}\}$, where the sum remains bounded as $N \to \infty$. Therefore

$$\left|\tilde{X}_0(N)(\mathbf{F}_{p^2})\right| = (p - 1)g_0 + O(1)$$

as $N \to \infty$. This is just (4.7) in Theorem 4.6. ∎

EXAMPLE 4.4 We consider the Fermat curve

$$X^3 + Y^3 + Z^3 = 0. \tag{4.9}$$

The substitution

$$x = \frac{3Z}{X + Y}, \quad y = \frac{9}{2}\frac{X - Y}{X + Y} + \frac{1}{2}$$

transforms the curve (4.9) into

$$C : y^2 - y = x^3 - 7.$$

The elliptic curve C is called a Weierstrass model of (4.9). We consider the curve C over the finite field $\mathbf{F}_q$. As a moduli space, C is a model for the modular curve $\tilde{X}_0(27)$ (Serre). It is known that

$$\dim S_2(\Gamma_0(27)) = 1,$$

and its generator is given by

$$\eta(3z)^2\eta(9z)^2 = q\prod_{n=1}^{\infty}(1-q^{3n})^2(1-q^{9n})^2$$

$$= \sum_{n=1}^{\infty}a(n)q^n \quad (q = e^{2\pi i z}),$$

where $\eta(z)$ denotes the *Dedekind eta function* defined by

$$\eta(z) = q^{\frac{1}{24}}\prod_{n=1}^{\infty}(1-q^n).$$

Also we have

$$|C(\mathbf{F}_p)| = p + 1 - a(p) \quad (p \neq 3)$$

and

$$|C(\mathbf{F}_{2^l})| = 2^l + 1 - \alpha_1{}^l - \bar{\alpha}_1{}^l \quad (\alpha_1 = \sqrt{-2}).$$

Now we put

$$\mathbf{F}_4 = \{0, 1, \alpha, \beta\}$$

with $\beta = 1 + \alpha = \alpha^2$. Then the points on C rational over $\mathbf{F}_4$ are given by the following table:

	P_1	P_2	P_3	P_4	P_5	P_6	P_7	P_8	Q
X	1	α	1	0	β	0	α	1	0
Y	0	0	1	α	0	β	1	α	1
Z	1	1	0	1	1	1	0	0	1

Let $D = \sum_{i=1}^{8} P_i$ and $G = 2Q$. Then, by Theorem 4.6, the modular curve code $C(D, G)$ for $\tilde{X}_0(27)$ has $n = 8$, $k = 2$ and the minimum distance d is at least 6 and one sees that $d = 6$ in this case.

REMARK 4.7 It is known that a modular curve $X_0(N)$ is of genus 1 if and only if

$$N \in \{11, 14, 15, 17, 19, 20, 21, 24, 27, 32, 36, 49\}.$$

In these cases, $X_0(N)$ is birational to an elliptic curve.

5.3   Asymptotic bounds

Let $C = C(P, G) = C(X, P, G)$ be an algebraic-geometric code, where X is a curve of genus g defined over $\mathbf{F}_q$, P is a set of $\mathbf{F}_q$-rational points on X of size n, and G is a divisor on X satisfying $2g - 2 < \deg(G) < n$. Then Theorem 4.6 tells us that the information rate R of C is $k/n = (\deg(G) - g + 1)/n$ and the relative minimum distance δ of C is $d/n \geq (n - \deg(G))/n$. Therefore, we have

$$R + \delta \geq 1 + \frac{1}{n} - \frac{g}{n}.$$

One way of thinking about the fact that we want both R and δ large while acknowledging that there is a trade off between these values is to say that we want $R + \delta$ large. For long codes, we consider the limit as n gets large. This means we consider a sequence of algebraic-geometric codes of increasing length. For this we need a sequence of curves X_i of genus g_i, a set of n_i rational points on X_i, and a chosen divisor G_i on X_i. Then we have

$$\lim_{n \to \infty} (R + \delta) \geq 1 - \lim_{i \to \infty} \frac{g_i}{n_i}.$$

Since we want $R + \delta$ to be big, we want $\lim_{n \to \infty} n/g$ to be as large as possible. Since $n \leq |X(\mathbf{F}_q)|$, we are prompted to make the definition, as in Section 4.3.2,

$$A(q) = \limsup_{g \to \infty} \frac{N_q(g)}{g}.$$

Suppose that we have a sequence of curves X_i defined over $\mathbf{F}_q$ satisfying

$$A(q) = \lim_{i \to \infty} \frac{N_i}{g_i},$$

where g_i is the genus of X_i and $N_i = |X_i(\mathbf{F}_q)|$. For each i, pick $Q_i \in X_i(\mathbf{F}_q)$, and set $P_i = X_i(\mathbf{F}_q) \setminus \{Q_i\}$. Also pick positive integers r_i with $2g_i - 2 < r_i < N_i - 1 = |P_i|$. Then the algebraic-geometric code $C_i = (X_i, P_i, r_i Q_i)$ has length $N_i - 1$, dimension $r_i + 1 - g_i$, and minimum distance at least $N_i - 1 - r_i$. If R_i is the information rate of C_i and δ_i is the relative minimum distance of C_i, then we have

$$R_i + \delta_i \geq 1 + \frac{1}{N_i - 1} - \frac{g_i}{N_i - 1}.$$

Put $R = \lim_{i \to \infty} R_i$ and $\delta = \lim_{i \to \infty} \delta_i$. Then we have

$$R + \delta \geq 1 - \frac{1}{A(q)}. \tag{4.10}$$

Now we introduce some new notations.

DEFINITION 4.7 Let q be a prime power. Then we define the following:

(1) Let n and d be positive integers with $d \leq n$. Then the quantity $A_q(n, d)$ is defined as the maximum value of M such that there is a linear code over $\mathbf{F}_q$ of length n with M codewords and minimum distance d. The Singleton bound (Theorem 1.5) means that $A_q(n, q) \leq q^{n-d+1}$.

(2) Let δ be a real number with $0 \leq \delta \leq 1$. Then we define

$$\alpha_q(\delta) = \limsup_{n \to \infty} \frac{1}{n} \log_q A_q(n, \delta n).$$

The quantity $\alpha_q(\delta)$ is the largest R such that there is a sequence of codes over $\mathbf{F}_q$ with relative minimum distance converging to δ and information rate converging to R.

(3) Put $\theta = 1 - 1/q$, and define the function H_q on the interval $0 \leq x \leq \theta$ by

$$H_q(x) = \begin{cases} 0 & \text{for } x = 0, \\ x \log_q(q - 1) - x \log_q x - (1 - x) \log_q(1 - x) & \text{for } 0 < x \leq \theta. \end{cases}$$

The function $H_q(x)$ is called the *Hilbert entropy function*.

With the notations as above, we have the following *Gilbert-Varshamov bound*: for $0 \leq \delta \leq \theta$,

$$\alpha_q(\delta) \geq 1 - H_q(\delta). \tag{4.11}$$

Furthermore, (4.10) means that

$$\alpha_q(\delta) \geq -\delta + 1 - \frac{1}{A(q)}.$$

Since the equation

$$R = -\delta + 1 - \frac{1}{A(q)}$$

defines a line of negative slope, it will intersect the *Gilbert-Varshamov curve*

$$R = 1 - H_q(\delta) \tag{4.12}$$

in either 0,1 or 2 points. If it intersects in two points, then we have an improvement on the Gilbert-Varshamov bound (4.11) in the interval between those two points.

From (4.6) we have the upper bound

$$A(q) \leq \sqrt{q} - 1$$

on $A(q)$. On the other hand, by Theorem 4.8 we have

$$A(q) = \sqrt{q} - 1$$

if $q = p^2$. Therefore,

$$\alpha_q(\delta) \geq -\delta + 1 - \frac{1}{\sqrt{q} - 1}. \tag{4.13}$$

It is not difficult to see that the *Tsfasman-Vladut-Zink line* (or *modular line*)

$$R = -\delta + 1 - \frac{1}{\sqrt{q} - 1}$$

and the Gilbert-Varshamov curve (4.12) will intersect in exactly two points δ_1, δ_2 whenever $q \geq 49$. Therefore, for $p \geq 7$ and $q = p^2$, the Tsfasman-Vladut-Zink bound (4.13) gives an improvement on the Gilbert-Varshamov bound for the possible asymptotic parameter of codes over the field $\mathbf{F}_q$ (Figure 4.1).

REMARK 4.8 For any value of q which is not a square, the exact value of $A(q)$ is unknown. The following conjecture is due to Manin:

Conjecture.
$$A(p^5) = p^2 - 1.$$

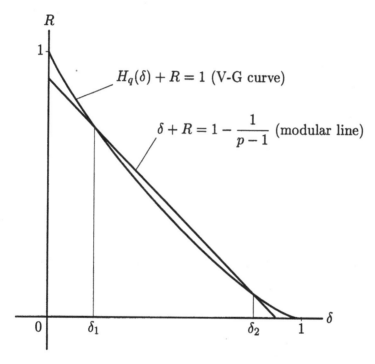

Figure 4.1

Chapter 5

THETA FUNCTIONS AND SELF-DUAL CODES

There is a connection between binary linear codes and lattices in n-space. To every lattice is associated its theta series. The Poisson summation formula gives a relation between the theta series of a lattice and that of its dual. For even unimodular lattices this tells us that the theta series is a modular form. Classical relations among modular forms then yield results about lattices and the associated linear codes. In this chapter we shall give an introduction to this fascinating relationship between seemingly disparate topics. In particular, we will give another proof for the MacWilliams identity. Our exposition in this chapter is based on [7]. A wealth of information and inspiration can be found in [4].

1. Lattices and codes

1.1 Lattices

DEFINITION 5.1 A *lattice* in n-space is a free abelian group Λ which is generated by a vector space basis of $\mathbf{R}^n$. Thus

$$\Lambda = \mathbf{Z}v_1 + \ldots + \mathbf{Z}v_n = \{a_1v_1 + \ldots + a_nv_n \ : \ a_i \in \mathbf{Z} \quad \text{for} \quad i = 1, \ldots, n\},$$

where $v_1, \ldots, v_n$ is a basis of $\mathbf{R}^n$. We call this basis also a *basis of the lattice* Λ.

The simplest example of a lattice is the set $\mathbf{Z}^n$ of all integral points in $\mathbf{R}^n$; it is generated by the standard basis of $\mathbf{R}^n$. The set $\mathbf{Z} + \mathbf{Z}\sqrt{2} = \{a + b\sqrt{2} : a, b \in \mathbf{Z}\}$ is a free abelian group of rank 2, but it is not a lattice.

We call

$$P = \{\lambda_1 v_1 + \ldots + \lambda_n v_n : 0 \leq \lambda_i \leq 1 \quad \text{for} \quad i = 1, \ldots, n\}$$

the *fundamental parallelotope* of the lattice $\Lambda = \mathbf{Z}v_1 + \ldots + \mathbf{Z}v_n$ with respect to the basis $v_1, \ldots, v_n$.

We think of $\mathbf{R}^n$ as a Euclidean vector space, equipped with the standard scalar product $x \cdot y = x_1 y_1 + \ldots + x_n y_n$ for $x = (x_1, \ldots, x_n)$ and $y = (y_1, \ldots, y_n)$ in $\mathbf{R}^n$. We will write x^2 instead of $x \cdot x$. The volume of P is

$$\text{vol}(P) = |\det(v_1, \ldots, v_n)|.$$

Any other basis of Λ is given by $(v_1, \ldots, v_n) \cdot A$ with a matrix $A \in GL_n(\mathbf{Z})$. Therefore the volume of P does not depend on the choice of the basis of Λ, and we can call $\text{vol}(\Lambda) = \text{vol}(P)$ the *volume of the lattice* Λ.

Of course we can work with any finite-dimensional Euclidean vector space instead of $\mathbf{R}^n$, as well. In this way we can, in particular, explain the concept of a lattice in a vector subspace of $\mathbf{R}^n$.

PROPOSITION 5.1 Let Λ' and Λ be lattices in $\mathbf{R}^n$ such that $\Lambda' \subseteq \Lambda$. Then Λ' is a subgroup of finite index in Λ, and

$$\text{vol}(\Lambda') = \text{vol}(\Lambda) \cdot [\Lambda : \Lambda'].$$

Proof. We indicate the ideas and leave the details to the reader. Let the column vectors $v_1, \ldots, v_n$ and $v'_1, \ldots, v'_n$ be bases of the lattices Λ and Λ', respectively. From $\Lambda' \subseteq \Lambda$ we obtain that $(v'_1, \ldots, v'_n) = (v_1, \ldots, v_n) \cdot A$ with some integral non-singular matrix A. We put $d = |\det(A)|$. Then $(dv_1, \ldots, dv_n) = (v'_1, \ldots, v'_n) \cdot dA^{-1}$. This shows that $d\Lambda \subseteq \Lambda'$ and that Λ' has finite index in Λ. Now it suffices to show that $[\Lambda : \Lambda'] = d$. We can choose the basis $v'_1, \ldots, v'_n$ in a special way such that $A = (a_{ij})$

is an upper triangular matrix with positive diagonal entries a_{jj} and $d = a_{11} \cdot \ldots \cdot a_{nn}$. It is easy to see then that the vectors $c_1 v_1 + \ldots + c_n v_n$ with $0 \leq c_j < a_{jj}$ form a set of coset representatives for Λ' in Λ. ∎

DEFINITION 5.2 Let Λ be a lattice in $\mathbf{R}^n$. Then we call

$$\Lambda^* = \{x \in \mathbf{R}^n : x \cdot y \in \mathbf{Z} \quad \text{for all} \quad y \in \Lambda\}$$

the *dual lattice* of Λ.

Let $v_1, \ldots, v_n$ be a lattice basis for Λ, and let $v_1^*, \ldots, v_n^*$ be the dual basis of $\mathbf{R}^n$. It is defined by $v_i \cdot v_j^* = 1$ for $i = j$ and $v_i \cdot v_j^* = 0$ otherwise. Then clearly $\Lambda^* = \mathbf{Z} v_1^* + \ldots + \mathbf{Z} v_n^*$. Therefore, Λ^* is indeed a lattice, and $v_1^*, \ldots, v_n^*$ is a lattice basis for Λ^*.

PROPOSITION 5.2 For any lattice Λ in $\mathbf{R}^n$ and its dual lattice Λ^* we have

$$\text{vol}(\Lambda) \cdot \text{vol}(\Lambda^*) = 1.$$

Proof. Let $v_1, \ldots, v_n$ be a basis for the lattice Λ, and let $v_1^*, \ldots, v_n^*$ be the dual basis. For the matrices $B = (v_1, \ldots, v_n)$ and $B^* = (v_1^*, \ldots, v_n^*)$ we have

$$B^T B^* = (v_i \cdot v_j^*)_{i,j} = I_n,$$

the n by n identity matrix. This yields

$$\text{vol}(\Lambda)\text{vol}(\Lambda^*) = |\det(B)\det(B^*)| = |\det(B^T B^*)| = \det(I_n) = 1. \quad ∎$$

DEFINITION 5.3 A lattice Λ in $\mathbf{R}^n$ is called an *integral lattice* if $\Lambda \subseteq \Lambda^*$. This means that $x \cdot y \in \mathbf{Z}$ holds for all $x, y \in \Lambda$. If $v_1, \ldots, v_n$ is a basis of the integral lattice Λ, then $A = (v_i \cdot v_j)_{i,j}$ is an integral matrix, and we call

$$\text{disc}(\Lambda) = \det(A)$$

the *discriminant* of the lattice Λ. From Propositions 5.1 and 5.2 it follows that $\mathrm{vol}(\Lambda)^2 = \det(A) = \mathrm{disc}(\Lambda)$ and

$$[\Lambda^* : \Lambda] = \frac{\mathrm{vol}(\Lambda)}{\mathrm{vol}(\Lambda^*)} = \mathrm{vol}(\Lambda)^2 = \mathrm{disc}(\Lambda)$$

for any integral lattice Λ. The matrix A is also called a *Gram matrix* for the lattice Λ.

We call Λ a *unimodular lattice* if $\Lambda = \Lambda^*$. Equivalently, Λ is integral and $\mathrm{disc}(\Lambda) = 1$.

A lattice Λ is called an *even lattice* if it is an integral lattice and $v \cdot v$ is even for all $v \in \Lambda$. For the integral matrix A introduced above this means that A has even entries in the diagonal.

Even unimodular lattices are of utmost interest. Their existence is not at all trivial. In contrast, it is easy to exhibit unimodular lattices; an example is $\mathbf{Z}^n$.

DEFINITION 5.4 Let Λ be a lattice in $\mathbf{R}^n$. Any $\mathbf{Z}$-submodule L of Λ is called a *sublattice* of Λ. Observe that the rank k of L may be smaller than n. In that case, L is not a lattice in $\mathbf{R}^n$, according to Definition 5.1, but L is a lattice in some subspace W of $\mathbf{R}^n$ which is isomorphic to $\mathbf{R}^k$. A sublattice L of Λ is called a *primitive sublattice* if the factor group Λ/L is a free $\mathbf{Z}$-module. If M is any subset of Λ, we call the $\mathbf{Z}$-submodule

$$M^\perp = \{v \in \Lambda \; : \; x \cdot v = 0 \quad \text{for all} \quad x \in M\}$$

the sublattice *orthogonal* to M.

Let $L_1, \ldots, L_m$ be sublattices of a lattice Λ in $\mathbf{R}^n$. We call Λ the *orthogonal direct sum* of $L_1, \ldots, L_m$, and we write $\Lambda = L_1 \oplus \ldots \oplus L_m$, if Λ as a $\mathbf{Z}$-module is the direct sum of the $\mathbf{Z}$-submodules $L_1, \ldots, L_m$ and if, moreover, we have $x \cdot y = 0$ for all $x \in \Lambda_i$, $y \in \Lambda_j$, and $i \neq j$.

PROPOSITION 5.3 Let Λ be a unimodular lattice in $\mathbf{R}^n$ and L a primitive sublattice of Λ. Then $\operatorname{disc}(L) = \operatorname{disc}(L^\perp)$.

Proof. Since Λ/L is a free $\mathbf{Z}$-module, there is a basis $v_1, \ldots, v_k$ of L which can be extended to a basis $v_1, \ldots, v_k, \ldots, v_n$ of Λ. Since $\Lambda^* = \Lambda$, the dual basis $v_1^*, \ldots, v_n^*$ is also a basis of Λ. A basis of $L^\perp$ is $v_{k+1}^*, \ldots, v_n^*$. Therefore, $v_{k+1}, \ldots, v_n$ is a basis of $(L^\perp)^*$. Thus we obtain

$$\Lambda = L + (L^\perp)^*, \qquad L \cap (L^\perp)^* = \{0\},$$

$$\Lambda = \Lambda^* = (L + (L^\perp)^*)^* = L^* + L^\perp, \qquad L^* \cap L^\perp = \{0\}.$$

So Λ is the direct sum of L and $(L^\perp)^*$, and also the direct sum of L^* and $L^\perp$. Thus we get

$$
\begin{aligned}
\operatorname{disc}(L^\perp) &= [(L^\perp)^* : L^\perp] = [L + (L^\perp)^* : L + L^\perp] \\
&= [\Lambda : L + L^\perp] = [L^* + L^\perp : L + L^\perp] = [L^* : L] = \operatorname{disc}(L).
\end{aligned}
$$

∎

1.2 Constructing lattices from binary codes

DEFINITION 5.5 We denote by $\rho : \mathbf{Z}^n \to \mathbf{F}_2^n$ the reduction map modulo 2. Let C be a binary $[n, k]$-code, i. e., C is a subspace of dimension k in the vector space $\mathbf{F}_2^n$. It is easy to see that the preimage $\rho^{-1}(C) \subseteq \mathbf{Z}^n$ is a lattice in $\mathbf{R}^n$: Because of $\mathbf{F}_2^n/C \simeq \mathbf{F}_2^{n-k}$, the preimage $\rho^{-1}(C)$ is a subgroup with finite index 2^{n-k} in $\mathbf{Z}^n$. Thus it is a free abelian group of rank n. And if $e_1, \ldots, e_n$ denote the standard unit vectors in $\mathbf{R}^n$, then the preimages $2e_1, \ldots, 2e_n$ of 0 belong to $\rho^{-1}(C)$ and form a vector space basis of $\mathbf{R}^n$. We call

$$\Lambda_C = \frac{1}{\sqrt{2}} \cdot \rho^{-1}(C)$$

the *lattice associated with the code* C. Its volume is

$$\operatorname{vol}(\Lambda_C) = 2^{-n/2}\operatorname{vol}(\rho^{-1}(C)) = 2^{-n/2}[\mathbf{F}_2^n : C] = 2^{\frac{n}{2}-k}.$$

The factor $1/\sqrt{2}$ in the definition of Λ_C seems to be artificial at first sight. It is justified by the simplicity of the following result, and it will further be justified by the construction of an even unimodular lattice in $\mathbf{R}^8$.

A code is called *doubly-even* if the weight of every codeword is a multiple of 4.

PROPOSITION 5.4 Let Λ_C be the lattice associated with a binary linear code C. Then we have:

(1) The code C is self-orthogonal if and only if Λ_C is an integral lattice.

(2) The code C is doubly-even if and only if Λ_C is an even lattice.

(3) The code C is self-dual if and only if Λ_C is unimodular.

Proof. Let $C \subseteq \mathbf{F}_2^n$ be a binary $[n, k]$-code. For $c \in \mathbf{F}_2^n$ we denote by $\dot{c}$ the unique vector in $\rho^{-1}(C)$ whose components are 0 or 1. Then any vector v in the lattice Λ_C can uniquely be written as $v = (\dot{c} + 2z)/\sqrt{2}$ where $c \in C$ and $z \in \mathbf{Z}^n$. For v and $w = (\dot{c}' + 2z')/\sqrt{2}$ with $c' \in C$, $z' \in \mathbf{Z}^n$ we obtain

$$v^2 = \frac{1}{2}(\dot{c}^2 + 4\dot{c} \cdot z + 4z^2),$$

$$v \cdot w = \frac{1}{2}((v+w)^2 - v^2 - w^2) \equiv \frac{1}{4}((\dot{c}+\dot{c}')^2 - \dot{c}^2 - (\dot{c}')^2) \equiv \frac{1}{2}\dot{c} \cdot \dot{c}' \bmod \mathbf{Z}.$$

These relations show that Λ_C is an integral lattice if and only if $\dot{c} \cdot \dot{c}'$ is even for all $c, c' \in C$. This means that $c \cdot c' = 0$ for all $c, c' \in C$, which by definition is equivalent to $C \subseteq C^\perp$. Thus we have proved assertion (1).

In the same way we see that $v^2 \in 2\mathbf{Z}$ for all $v \in \Lambda_C$ if and only if $\dot{c}^2 \in 4\mathbf{Z}$ for all $c \in C$. But $\dot{c}^2$ is the weight of c. This proves (2). The discriminant

$$\text{disc}(\Lambda_C) = \text{vol}(\Lambda_C)^2 = 2^{n-2k}$$

is equal to 1 if and only if $n = 2k$, i. e., if and only if $\dim(C^\perp) = n - k = k = \dim(C)$. Together with $C \subseteq C^\perp$, this proves (3). ∎

PROPOSITION 5.5 Every doubly-even binary linear code C is self-orthogonal. If C is self-orthogonal and $\dim(C) = \dim(C^\perp)$, then C is self-dual.

Proof. The first assertion follows from Proposition 5.4. Then the second one is obvious. ∎

DEFINITION 5.6 For any code $C \subseteq F_q^n$, the *extended code* $\tilde{C} \subseteq F_q^{n+1}$ consists of all words $(c_1, \ldots, c_{n+1}) \in F_q^{n+1}$ such that $(c_1, \ldots, c_n) \in C$ and $c_1 + \ldots + c_n + c_{n+1} = 0$. Clearly, if C is a linear $[n, k]$-code, then $\tilde{C}$ is a linear $[n+1, k]$-code.

1.3   Examples

EXAMPLE 5.1 The *Hamming code* is the code H in F_2^7 with the generator matrix

$$G = \begin{pmatrix} 1 & 0 & 0 & 0 & 1 & 1 & 1 \\ 0 & 1 & 0 & 0 & 0 & 1 & 1 \\ 0 & 0 & 1 & 0 & 1 & 0 & 1 \\ 0 & 0 & 0 & 1 & 1 & 1 & 0 \end{pmatrix}$$

in standard form. According to Proposition 1.3, a parity check matrix for H is

$$\begin{pmatrix} 1 & 0 & 1 & 1 & 1 & 0 & 0 \\ 1 & 1 & 0 & 1 & 0 & 1 & 0 \\ 1 & 1 & 1 & 0 & 0 & 0 & 1 \end{pmatrix}.$$

This code was invented and used in 1948 by R. W. Hamming (1915 - 1998). The code H has minimum distance 3. So it can correct one error, with an information rate of 4/7. The 16 codewords can easily be listed. This yields the weight enumerator

$$W_H(X, Y) = X^7 + 7X^4Y^3 + 7X^3Y^4 + Y^7.$$

The extended Hamming code $\tilde{H} \subseteq F_2^8$ is obtained by adding a parity check bit to every codeword of H. So a generator matrix of $\tilde{H}$ in standard form is

$$\begin{pmatrix} 1 & 0 & 0 & 0 & 1 & 1 & 1 & 0 \\ 0 & 1 & 0 & 0 & 0 & 1 & 1 & 1 \\ 0 & 0 & 1 & 0 & 1 & 0 & 1 & 1 \\ 0 & 0 & 0 & 1 & 1 & 1 & 0 & 1 \end{pmatrix}.$$

For the weight enumerator of $\tilde{H}$ we obtain

$$W_{\tilde{H}}(X,Y) = X^8 + 14X^4Y^4 + Y^8.$$

This shows that $\tilde{H}$ is doubly-even, and from Proposition 5.5 it follows that $\tilde{H}$ is self-dual. Now Proposition 5.4 tells us that the corresponding lattice $\Lambda_{\tilde{H}}$ is an even unimodular lattice in $\mathbf{R}^8$. This lattice is usually called the E_8-*lattice*, and it is denoted by E_8.

For every lattice basis $v_1, \ldots, v_8$ of the even unimodular lattice E_8, the Gram matrix $A = (v_i \cdot v_j)_{i,j}$ is integral with even diagonal entries and determinant 1. We shall give a specific basis for which all diagonal entries of A are equal to 2, i. e., for which every basis vector v_i has squared length $v_i^2 = 2$. We start with 7 vectors $f_1, \ldots, f_7$ in E_8 which come from preimages of the 7 codewords of weight 7 in H and are given by

$$\begin{aligned} \sqrt{2}f_1 &= (0,0,1,0,1,0,1,1), \\ \sqrt{2}f_2 &= (1,0,1,0,0,1,0,1), \\ \sqrt{2}f_3 &= (0,0,0,1,1,1,0,1), \\ \sqrt{2}f_4 &= (1,1,0,0,1,0,0,1), \\ \sqrt{2}f_5 &= (0,1,0,0,0,1,1,1), \\ \sqrt{2}f_6 &= (1,0,0,1,0,0,1,1), \\ \sqrt{2}f_7 &= (0,1,1,1,0,0,0,1). \end{aligned}$$

We have $f_i^2 = 2$ for $i = 1, \ldots, 7$. For every pair (i,j) with $i \neq j$ there are exactly two positions for which f_i and f_j both have coordinates $\neq 0$ in these positions. Therefore, $f_i \cdot f_j = 1$ for $i \neq j$. Now we put

$$v_1 = f_1, \qquad v_i = f_i - f_{i-1} \quad \text{for} \quad 2 \leq i \leq 7.$$

The formulas for $f_i \cdot f_j$ imply that

$$
\begin{aligned}
v_i^2 &= 2 \quad \text{for} \quad i = 1, \ldots, 7, \\
v_i \cdot v_{i+1} &= v_{i+1} \cdot v_i = -1 \quad \text{for} \quad i = 1, \ldots, 6, \\
v_i \cdot v_j &= 0 \quad \text{otherwise.}
\end{aligned}
$$

One basis vector for E_8 is still missing. We choose

$$
v_8 = \frac{1}{\sqrt{2}}(0, -1, 0, -1, 1, 0, -1, 0)
$$

which comes from a preimage of one of the codewords of weight 4 in H. We observe that $v_8^2 = 2$, $v_8 \cdot f_i = 0$ for $i \in \{1, 2, 3, 4\}$, $v_8 \cdot f_i = -1$ for $i \in \{5, 6, 7\}$,

$$
\begin{aligned}
v_8 \cdot v_i &= 0 \quad \text{for} \quad 1 \leq i \leq 7, \ i \neq 5, \\
v_8 \cdot v_5 &= -1.
\end{aligned}
$$

Thus we have

$$
A = (v_i \cdot v_j)_{i,j} = \begin{pmatrix}
2 & -1 & 0 & 0 & 0 & 0 & 0 & 0 \\
-1 & 2 & -1 & 0 & 0 & 0 & 0 & 0 \\
0 & -1 & 2 & -1 & 0 & 0 & 0 & 0 \\
0 & 0 & -1 & 2 & -1 & 0 & 0 & 0 \\
0 & 0 & 0 & -1 & 2 & -1 & 0 & -1 \\
0 & 0 & 0 & 0 & -1 & 2 & -1 & 0 \\
0 & 0 & 0 & 0 & 0 & -1 & 2 & 0 \\
0 & 0 & 0 & 0 & -1 & 0 & 0 & 2
\end{pmatrix}.
$$

The determinant of A can be computed by expanding it with respect to rows: Let A_k be the matrix which consists of the lowest k rows and columns of A. Then we get

$$
\begin{aligned}
\det(A) &= 2\det(A_7) - \det(A_6) = 3\det(A_6) - 2\det(A_5) \\
&= \ldots = 5\det(A_4) - 4\det(A_3) = 25 - 24 = 1.
\end{aligned}
$$

Since E_8 is unimodular, this shows us that $v_1, \ldots, v_8$ is indeed a lattice basis and that A is a Gram matrix for E_8.

The entries in A tell us which angle is enclosed between any of the basis vectors: The vectors v_i and v_j are perpendicular if $v_i \cdot v_j = 0$, and the enclosed angle is $2\pi/3$ if $v_i \cdot v_j = -1$. This information about the angles is visualized in a condensed form in the so-called *Coxeter-Dynkin diagram* of the vectors $v_1, \ldots, v_8$. By definition, this is the graph with 8 vertices corresponding to the vectors v_i, and two vertices v_i and v_j are connected by an edge if and only if $v_i \cdot v_j = -1$. Thus the entries in A give us the following graph.

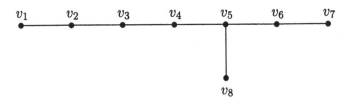

We have also shown that E_8 is a root lattice. This concept is defined as follows.

DEFINITION 5.7 Let Λ be an even lattice in $\mathbf{R}^n$. A vector $x \in \Lambda$ is called a *root* in Λ if $x^2 = 2$. The lattice Λ is called a *root lattice* if Λ is generated by the set of its roots.

A famous theorem of E. Witt (1941) states that every root lattice Λ in $\mathbf{R}^n$ has a basis $v_1, \ldots, v_n$ such that every v_i is a root and $v_i \cdot v_j \in \{0, -1\}$ for $1 \leq i < j \leq n$. A proof of this theorem, together with a classification of all root lattices, can be found in [7], Chapter 1.

Example 5.1, continued. Since E_8 is an even lattice, x^2 is a non-negative even integer for every $x \in E_8$. In particular, the shortest non-zero vectors in E_8 are the roots of E_8. It is interesting to calculate the number of roots in E_8. More generally, we let

$$a_n = |\{x \in E_8 \ : \ x^2 = 2n\}|$$

be the number of vectors in E_8 with squared length equal to $2n$. With these numbers as coefficients, we form the Fourier series

$$\Theta(z, E_8) = \sum_{n=0}^{\infty} a_n e^{2\pi i n z} = \sum_{x \in E_8} e^{\pi i x^2 z}$$

which is called the *theta function* of the lattice E_8. In section 5.2 we will introduce the theta function of an arbitrary lattice, and we will establish a transformation formula which shows that $\Theta(z, E_8)$ is a modular form of weight 4 on the full modular group Γ.

Obviously we have $a_0 = 1$. It is easy to see that $a_1 = 240$:

PROPOSITION 5.6 Let E_8 be the lattice in $\mathbf{R}^8$ corresponding to the extended Hamming code $\tilde{H}$. Then E_8 is an even unimodular lattice, it is a root lattice, and the number of its roots is 240.

Proof. For codewords $c \in \tilde{H}$ we use the notation $\dot{c} \in \mathbf{Z}^8$ from the proof of Proposition 5.4, and we write $x \in E_8$ uniquely as $x = (\dot{c} + 2b)/\sqrt{2}$ with $c \in \tilde{H}$, $b \in \mathbf{Z}^8$. We suppose that $x^2 = 2$, whence $(\dot{c} + 2b)^2 = 4$. If $c = 0$, then we have $b^2 = 1$, and this gives us exactly 16 roots x in E_8. The codeword $c = (1, \ldots, 1)$ does not give roots of E_8. Now let $c = (c_1, \ldots, c_8)$ be one of the 14 codewords of weight 4. Then for $b = (b_1, \ldots, b_8)$ we must have $b_i = 0$ whenever $c_i = 0$, and we can choose $b_i \in \{0, -1\}$ if $c_i = 1$. This gives us exactly $2^4 \cdot 14$ roots in E_8. Altogether we obtain $a_1 = 16 + 16 \cdot 14 = 240$. ∎

EXAMPLE 5.2 We consider the dual code $H^\perp$ of the Hamming code H. A generator matrix for $H^\perp$ is the parity check matrix of H which was shown at the beginning of Example 5.1. So from the self-duality of the extended Hamming code it follows that

$$H^\perp = \{(c_1, \ldots, c_7) \in \mathbf{F}_2{}^7 \; : \; (c_1, \ldots, c_7, 0) \in \tilde{H}\},$$

and that the codewords in $H^\perp$ are, besides of 0, exactly the 7 codewords of H with weight 4. The corresponding lattice $\Lambda_{H^\perp} = (1/\sqrt{2})\rho^{-1}(H^\perp)$ in $\mathbf{R}^7$ is usually called the E_7-lattice, and it is denoted by E_7. We have

$$E_7 = \{(x_1, \ldots, x_7) \in \mathbf{R}^7 \; : \; (x_1, \ldots, x_7, 0) \in E_8\}.$$

When we identify E_7 with its embedding in $\mathbf{R}^8$, then E_7 is a sublattice in E_8. The factor group $E_8/E_7 \simeq \sqrt{2}\mathbf{Z}$ is the one-dimensional lattice which is represented by the sublattice of E_8 with basis vector

$(1/\sqrt{2})(0,0,0,0,0,0,0,2) \in E_8$. Therefore, E_7 is a primitive sublattice of E_8 with $E_7^{\perp} \simeq \sqrt{2}\mathbf{Z}$, and from Proposition 5.3 we obtain

$$\text{disc}(E_7) = \text{disc}(\sqrt{2}\mathbf{Z}) = \text{vol}(\sqrt{2}\mathbf{Z})^2 = 2.$$

This helps us to find a lattice basis for E_7: Let $v_1,\ldots,v_8$ be the basis of E_8 in Example 5.1. The vectors $v_1,\ldots,v_7$ have 0 as their last coordinate. If we drop this coordinate then we get 7 linearly independent vectors in E_7. The matrix of their inner products is obtained by dropping the first row and column from the matrix A in Example 5.1, i. e., we obtain the matrix

$$A_7 = \begin{pmatrix} 2 & -1 & 0 & 0 & 0 & 0 & 0 \\ -1 & 2 & -1 & 0 & 0 & 0 & 0 \\ 0 & -1 & 2 & -1 & 0 & 0 & 0 \\ 0 & 0 & -1 & 2 & -1 & 0 & -1 \\ 0 & 0 & 0 & -1 & 2 & -1 & 0 \\ 0 & 0 & 0 & 0 & -1 & 2 & 0 \\ 0 & 0 & 0 & -1 & 0 & 0 & 2 \end{pmatrix}.$$

The same computation as above shows that $\det(A_7) = 2 = \text{disc}(E_7)$. Therefore, our 7 vectors form a basis for E_7. Now it is easy to establish the following result.

PROPOSITION 5.7 Let E_7 be the lattice in $\mathbf{R}^7$ associated with the dual $H^{\perp}$ of the Hamming code H. Then E_7 is an even lattice with discriminant 2, and it is a root lattice with Coxeter-Dynkin diagram

There are exactly 126 roots in E_7.

Proof. We have found a basis of E_7 consisting of roots and whose Gram matrix is A_7. From A_7 we read off the Coxeter-Dynkin diagram. The roots in E_7 are obtained in the same way as in Proposition 5.6. So we get $2 \cdot 7$ roots from preimages of the codeword 0 in $H^\perp$, and we get 2^4 roots from each of the 7 codewords of weight 4 in $H^\perp$. Thus the total number of roots is $14 + 7 \cdot 2^4 = 126$. ∎

EXAMPLE 5.3 Let D_n denote the set of all $x = (x_1, \ldots, x_n)$ in $\mathbf{Z}^n$ such that $x_1 + \ldots + x_n$ is even. Then D_n is a lattice in $\mathbf{R}^n$. It is called the *checkerboard lattice*, for obvious reasons. Clearly, D_n is a sublattice in $\mathbf{Z}^n$ with index $[\mathbf{Z}^n : D_n] = 2$. Therefore, from Proposition 5.1 and Definition 5.3 we get $\text{vol}(D_n) = 2$ and $\text{disc}(D_n) = 4$. We denote by $e_1, \ldots, e_n$ the standard unit vectors in $\mathbf{R}^n$. For $n \geq 2$ we put

$$v_1 = e_1 + e_2, \qquad v_k = e_k - e_{k-1} \quad \text{for} \quad 2 \leq k \leq n.$$

The vectors $v_1, \ldots, v_n$ belong to D_n. It is straightforward to compute $\det(v_1, \ldots, v_n) = 2 = \text{vol}(D_n)$. It follows that $v_1, \ldots, v_n$ is a lattice basis for D_n. Since $v_k^2 = 2$, we observe that D_n is a root lattice. The reader may easily compute the Gram matrix and draw the Coxeter-Dynkin diagram of the basis $v_1, \ldots, v_n$.

We are interested in the dual lattice D_n^*. From $D_n \subseteq \mathbf{Z}^n$ we obtain $D_n^* \supseteq (\mathbf{Z}^n)^* = \mathbf{Z}^n \supseteq D_n$. Thus D_n is an integral lattice and

$$[D_n^* : D_n] = \text{disc}(D_n) = 4.$$

For $x = (x_1, \ldots, x_n)$ in D_n we have $x^2 = x_1^2 + \ldots + x_n^2 \equiv x_1 + \ldots + x_n \equiv 0 \pmod 2$. Thus D_n is an even lattice. The vectors e_1 and $v = (1/2)(e_1 + \ldots + e_n) = (1/2, \ldots, 1/2)$ belong to D_n^*, but not to D_n. We have $2e_1 \in D_n$, $4v \in D_n$, and $2v$ belongs to D_n if and only if n is even. This shows that the factor group D_n^*/D_n is generated by the residue classes of e_1 and v mod D_n, that D_n^*/D_n is cyclic and generated by v mod D_n if n is odd, and that D_n^*/D_n is the Kleinian four-group if n is even.

Now we suppose that $n \equiv 0 \pmod 4$. We follow an idea of Serre ([31], p. 51) to construct a unimodular lattice Γ_n in-between D_n and D_n^* as follows. Let Γ_n be the subgroup of $\mathbf{R}^n$ which is generated by D_n

and the vector $v = (1/2)(e_1 + \ldots + e_n) = (1/2, \ldots, 1/2)$. Obviously, Γ_n is a lattice satisfying $D_n \subsetneq \Gamma_n \subsetneq D_n^*$. From $2v \in D_n$ it follows that $[\Gamma_n : D_n] = 2$. Because of $v^2 = n/4 \in \mathbf{Z}$, Γ_n is an integral lattice. We obtain

$$\text{disc}(\Gamma_n) = \text{vol}(\Gamma_n)^2 = \frac{\text{disc}(D_n)}{[\Gamma_n : D_n]^2} = 1.$$

Thus Γ_n is unimodular. If we suppose, moreover, that $n \equiv 0 (\text{mod } 8)$, then v^2 is even and Γ_n is an even unimodular lattice.

So in $\mathbf{R}^8$ we have the even unimodular lattices Γ_8 and E_8, and in $\mathbf{R}^{16}$ we have the even unimodular lattices Γ_{16} and $E_8 \oplus E_8$, the orthogonal direct sum of two copies of E_8. The question arises if there are any relations among these lattices. Partial answers will be given below in this subsection.

DEFINITION 5.8 Two lattices Λ_1 and Λ_2 in $\mathbf{R}^n$ are called *isomorphic lattices* if there is an orthogonal transformation S of $\mathbf{R}^n$ such that $S(\Lambda_1) = \Lambda_2$. If the lattice Λ_1 is integral, or even, or unimodular, then any isomorphic lattice has the same property. An *automorphism* of the lattice Λ_1 is an orthogonal transformation S of $\mathbf{R}^n$ such that $S(\Lambda_1) = \Lambda_1$. Clearly, the automorphisms of Λ_1 form a group.

It can be deduced from Theorem 5.14 in Subsection 5.2.1 and from a classification of root lattices that every even unimodular lattice in $\mathbf{R}^8$ is isomorphic with E_8. So Γ_8 and E_8 are isomorphic. But we can see at once that Γ_{16} and $E_8 \oplus E_8$ are not isomorphic. We need some knowledge about roots in D_n and in Γ_n:

PROPOSITION 5.8 For any integer $n \geq 2$, there are exactly $2n(n-1)$ roots in the lattice D_n. If $n \equiv 0 (\text{mod } 4)$, $n \neq 8$, then every root in Γ_n belongs to D_n, and Γ_n is not a root lattice. The number of roots in Γ_8 is 240.

Proof. By definition, the roots in D_n are the vectors $x = (x_1, \ldots, x_n)$ in $\mathbf{Z}^n$ for which $x_1 + \ldots + x_n$ is even and $x^2 = x_1^2 + \ldots + x_n^2 = 2$. Thus the roots are $\pm e_i \pm e_j$ where $1 \le i < j \le n$ and the signs are arbitrary. Their number is $2^2 \cdot n(n-1)/2 = 2n(n-1)$.

Let $n \equiv 0 \pmod 4$. Any vector $y \in \Gamma_n$, $y \notin D_n$, can be written as $y = v + x$ with $x \in D_n$ and $v = (1/2)(e_1 + \ldots + e_n)$. The condition that y is a root yields $\sum_{i=1}^n (2x_i + 1)^2 = 8$. This is possible only if $n = 8$ and $x_i \in \{0, -1\}$ for $i = 1, \ldots, n$. This proves the second assertion. Because of $x \in D_n$ we obtain exactly $(1/2) \cdot 2^8$ roots $y \in \Gamma_8$ with $y \notin D_8$. Thus the number of roots in Γ_8 is $16 \cdot 7 + 2^7 = 240$. ∎

Example 5.3, continued. By Proposition 5.8, the number of roots in Γ_{16} is $32 \cdot 15 = 480$. In $E_8 \oplus E_8$ we have $240 + 240 = 480$ roots, the same number as in Γ_{16}. But since $E_8 \oplus E_8$ is generated by its roots and Γ_{16} is not, these lattices cannot be isomorphic. In spite of this fact, in Section 5.2.1 we will be able to show that these lattices have identical theta series, i. e., that

$$|\{x \in \Gamma_{16} \,:\, x^2 = n\}| = |\{x \in E_8 \oplus E_8 \,:\, x^2 = n\}|$$

holds for all n.

As a final remark we mention that $D_n = \rho^{-1}(C)$ is the preimage of the binary $[n, n-1]$-code C of all words $(c_1, \ldots, c_n)$ in $\mathbf{F}_2^n$ such that $c_1 + \ldots + c_n = 0$. Of course, this does not mean that D_n is the lattice associated to C. It is not difficult to show that D_n is isomorphic to a lattice $(1/\sqrt{2})\rho^{-1}(C_n)$ associated to some code $C_n \subseteq \mathbf{F}_2^n$ if and only if n is even. If $n = 2m$, then the appropriate code C_n consists of all words $(c, c) \in \mathbf{F}_2^n$ such that $c = (c_1, \ldots, c_m) \in \mathbf{F}_2^m$ and $c_1 + \ldots + c_m = 0$. We omit the proof.

1.4 The lattices associated to a code and the dual code

In Section 5.2.2 we will need the following relation between the lattices associated to a code and that for the dual code.

PROPOSITION 5.9 For any binary linear code $C \subseteq \mathbf{F}_2^n$ we have

$$\Lambda_C^* = \Lambda_{C^\perp}.$$

Proof. Let $x \in \Lambda_C$ and $y \in \Lambda_{C^\perp}$ be given. Then $x = (1/\sqrt{2})v$, $y = (1/\sqrt{2})w$ for some $v \in \rho^{-1}(C)$, $w \in \rho^{-1}(C^\perp)$. It follows that $\rho(v) \cdot \rho(w) = 0$ in $\mathbf{F}_2$, hence $v \cdot w \in 2\mathbf{Z}$ and $x \cdot y \in \mathbf{Z}$. Since x and y are arbitrary, we have shown that $\Lambda_{C^\perp} \subseteq \Lambda_C^*$.

Now let $k = \dim(C)$, $\mu = \mathrm{vol}(\Lambda_C)$, $\mu^* = \mathrm{vol}(\Lambda_C^*)$. Proposition 5.2 gives us $\mu^* = \frac{1}{\mu}$. From $\dim(C^\perp) = n - k$ and the formula at the beginning of Section 5.1.2 we obtain $\mu = 2^{\frac{n}{2} - k}$ and

$$\mathrm{vol}(\Lambda_C^\perp) = 2^{\frac{n}{2} - \dim(C^\perp)} = 2^{k - \frac{n}{2}} = \frac{1}{\mu} = \mu^* = \mathrm{vol}(\Lambda_C^*).$$

Since $\Lambda_{C^\perp}$ is contained in Λ_C^* and the volumes are equal, it follows that the lattices are equal. ∎

2.   Theta functions and weight distributions

2.1   The theta function of a lattice

DEFINITION 5.9 Let Λ be a lattice in $\mathbf{R}^n$. The *theta function of the lattice* Λ is defined to be the series

$$\Theta_\Lambda(z) = \Theta(z, \Lambda) = \sum_{v \in \Lambda} e^{\pi i v^2 z},$$

where z is a complex variable. We will write $q = e^{2\pi i z}$ throughout this section. Then we have

$$\Theta(z, \Lambda) = \sum_{v \in \Lambda} q^{\frac{1}{2} v \cdot v}.$$

If Λ is an integral lattice then the theta function is a Fourier series

$$\Theta(z, \Lambda) = \sum_{m=0}^{\infty} a\left(\frac{m}{2}\right) q^{\frac{1}{2} m}$$

whose coefficients

$$a\left(\frac{m}{2}\right) = \left| \left\{ x \in \Lambda \ : \ v^2 = v \cdot v = \frac{m}{2} \right\} \right|$$

are the numbers of vectors in the lattice Λ with squared length $m/2$. If Λ is an even lattice then the theta function reads

$$\Theta(z, \Lambda) = \sum_{m=0}^{\infty} a(m) q^m.$$

We need to know where the series converges and defines an analytic function:

LEMMA 5.10 For any lattice Λ in $\mathbf{R}^n$, the theta function $\Theta(z, \Lambda)$ is holomorphic on the upper half-plane $\mathcal{H}$.

Proof. It suffices to show that the series converges absolutely uniformly on compact subsets of $\mathcal{H}$. If $v_1, \ldots, v_n$ is a basis of (column vectors) for the lattice Λ, then we have $\Lambda = B\mathbf{Z}^n$ with the real invertible matrix $B = (v_1, \ldots, v_n)$. On the unit sphere $x^2 = 1$ the function $x \mapsto (Bx)^2$ has a positive minimum ϵ. It follows that $(Bx)^2 \geq \epsilon x^2$ for all $x \in \mathbf{R}^n$. Let $\delta > 0$ be given. Then for $z \in \mathcal{H}$ with $\mathrm{Im}(z) \geq \delta$ we obtain

$$\sum_{v \in \Lambda} |q^{\frac{1}{2}v^2}| = \sum_{x \in \mathbf{Z}^n} |q^{\frac{1}{2}(Bx)^2}| \leq \sum_{x \in \mathbf{Z}^n} e^{-\pi \mathrm{Im}(z)\epsilon x^2}$$
$$\leq \sum_{x \in \mathbf{Z}^n} e^{-\pi \delta \epsilon x^2} = \left(\sum_{m=-\infty}^{\infty} e^{-\pi \delta \epsilon m^2} \right)^n.$$

The right hand side converges and does not depend on z. Thus the theta series of Λ converges absolutely uniformly on the half-plane $\mathrm{Im}(z) \geq \delta$ for every $\delta > 0$. ∎

EXAMPLE 5.4 The theta function of the lattice $\mathbf{Z}^n$ is

$$\Theta(z, \mathbf{Z}^n) = \theta(z)^n,$$

where

$$\theta(z) = \sum_{m=-\infty}^{\infty} e^{\pi i m^2 z} = \Theta(z, \mathbf{Z})$$

is Jacobi's theta function. It is well-known and will be deduced from Theorem 5.13 in this subsection that $\theta(z)$ is a modular form of weight $\frac{1}{2}$ on the theta group Γ_θ, which is the subgroup generated by $z \mapsto z+2$ and $z \mapsto -1/z$ in the full modular group Γ. Thus $\Theta(z, \mathbf{Z}^n)$ is a modular form of weight $\frac{n}{2}$ on Γ_θ. The Fourier coefficient $a(m) = a_n(m)$ of $\Theta(z, \mathbf{Z}^n)$ is the number of representations of m as a sum of n squares of integers. For small even values of n, i. e., for small integral weights of the modular form $\Theta(z, \mathbf{Z}^n)$, this modular form can be identified with Eisenstein series, and in this way one obtains Jacobi's famous formulas for $a_4(m)$, $a_6(m)$ and $a_8(m)$, namely,

$$a_4(m) = 8 \sum_{d\mid m,\, 4 \nmid d} d,$$

$$a_6(m) = 4 \sum_{d\mid m} \left(4\chi(\tfrac{m}{d}) - \chi(d)\right) d^2,$$

$$a_8(m) = 16 \sum_{d\mid m} (-1)^{m-d} d^3,$$

where $\chi(d) = (\frac{-1}{d})$ is the non-trivial character modulo 4. Also, Fermat's formula

$$a_2(m) = 4 \sum_{d\mid m} \chi(d)$$

can be obtained in this way. (See Chapter 9 in [8].)

The lattice $\mathbf{Z}^n$ is self-dual. For an arbitrary lattice Λ we will prove a relation between the theta function of Λ and that of the dual lattice Λ^*. As a tool we will need the Poisson summation formula.

THEOREM 5.11 (POISSON SUMMATION FORMULA) Let Λ be a lattice in $\mathbf{R}^n$ and Λ^* its dual lattice. Let f be a complex-valued continuous function on $\mathbf{R}^n$ with the following properties.

(1) The integral $\int_{\mathbf{R}^n} |f(x)|dx$ converges.

(2) The series $\sum_{v\in\Lambda} |f(v+u)|$ converges uniformly for u on every compact subset of $\mathbf{R}^n$.

(3) The series $\sum_{u\in\Lambda^*} |\hat{f}(u)|$ converges, where $\hat{f}(y) = \int_{\mathbf{R}^n} f(x)e^{-2\pi i x\cdot y}dx$ for $y \in \mathbf{R}^n$ denotes the Fourier transform of f.

Then we have

$$\sum_{v\in\Lambda} f(v) = \frac{1}{\mathrm{vol}(\Lambda)} \cdot \sum_{u\in\Lambda^*} \hat{f}(u).$$

Proof. Because of (2) the formula

$$F(u) = \sum_{v\in\Lambda} f(v+u)$$

defines a continuous function F on $\mathbf{R}^n$. First we deal with the lattice $\Lambda = \mathbf{Z}^n$. In that case, the function F has the period 1 in each variable, and hence it has a Fourier series

$$\sum_{y \in \mathbf{Z}^n} a(y)e^{2\pi i u \cdot y}$$

with coefficients

$$a(y) = \int_W F(t)e^{-2\pi i y \cdot t}dt,$$

where we integrate on the unit cube W in $\mathbf{R}^n$. Because of (1) we can interchange the summation and the integration, and this yields

$$a(y) = \sum_{v \in \mathbf{Z}^n} \int_W f(v+t)e^{-2\pi i y \cdot t}dt.$$

Since v and y belong to $\mathbf{Z}^n$, we can replace t by $v+t$ in the exponent. The translates $v+W$ of the unit cube cover the space $\mathbf{R}^n$, and so we get

$$a(y) = \int_{\mathbf{R}^n} f(t)e^{-2\pi i y \cdot t}dt = \hat{f}(y).$$

Now from (3) it follows that the Fourier series of $F(u)$ converges absolutely and uniformly and represents the function $F(u)$. Therefore,

$$\sum_{v \in \Lambda} f(v+u) = \sum_{y \in \mathbf{Z}^n} \hat{f}(y)e^{2\pi i u \cdot y}$$

for $u \in \mathbf{R}^n$. We put $u = 0$ and obtain the assertion for $\Lambda = \mathbf{Z}^n$, since $\mathrm{vol}(\mathbf{Z}^n) = 1$.

An arbitrary lattice Λ in $\mathbf{R}^n$ has the form $\Lambda = B\mathbf{Z}^n$ with an invertible matrix B whose columns are a basis of Λ. The proof of Proposition 5.2 yields

$$\Lambda^* = (B^T)^{-1}\mathbf{Z}^n$$

for the dual lattice Λ^*. We put $f_B(x) = f(Bx)$. Then f_B satisfies the conditions (1), (2), (3) as well as f. Therefore we get

$$\sum_{v \in \Lambda} f(v) = \sum_{x \in \mathbf{Z}^n} f(Bx) = \sum_{x \in \mathbf{Z}^n} f_B(x) = \sum_{y \in \mathbf{Z}^n} \hat{f}_B(y)$$

and

$$\hat{f}_B(y) \ = \ \int_{\mathbf{R}^n} f(Bt)e^{-2\pi i t \cdot y}dt = \frac{1}{|\det(B)|}\int_{\mathbf{R}^n} f(s)e^{-2\pi i (B^{-1}s)\cdot y}ds$$

$$= \ \frac{1}{\mathrm{vol}(\Lambda)}\int_{\mathbf{R}^n} f(s)e^{-2\pi i s\cdot (B^T)^{-1}y}ds$$

$$= \ \frac{1}{\mathrm{vol}(\Lambda)}\hat{f}\left((B^T)^{-1}y\right),$$

so finally

$$\sum_{v\in\Lambda} f(v) = \frac{1}{\mathrm{vol}(\Lambda)}\sum_{y\in\mathbf{Z}^n}\hat{f}\left((B^T)^{-1}y\right) = \frac{1}{\mathrm{vol}(\Lambda)}\sum_{u\in\Lambda^*}\hat{f}(u). \quad \blacksquare$$

Using Proposition 5.2, we can write Theorem 5.11 in a more symmetric form:

$$\sqrt{\mathrm{vol}(\Lambda)}\sum_{v\in\Lambda} f(v) = \sqrt{\mathrm{vol}(\Lambda^*)}\sum_{u\in\Lambda^*}\hat{f}(u).$$

For the application of Theorem 5.11 to the theta function of a lattice, we need the standard result on the Fourier transform of $e^{-\pi x^2}$:

LEMMA 5.12 The function $f(x) = e^{-\pi x^2}$ on $\mathbf{R}^n$ is equal to its own Fourier transform; for all $y \in \mathbf{R}^n$ we have

$$\int_{\mathbf{R}^n} e^{-\pi x^2}e^{-2\pi i x\cdot y}dx = e^{-\pi y^2}.$$

Proof. Because of

$$e^{-\pi x^2 - 2\pi i x\cdot y} = \prod_{k=1}^{n} e^{-\pi x_k^2 - 2\pi i x_k y_k}$$

and Fubini's theorem, it suffices to prove the formula for $n = 1$. We consider the derivative of the Fourier transform

$$\hat{f}(y) = \int_{\mathbf{R}} e^{-\pi x^2 - 2\pi i x y}dx$$

for $y \in \mathbf{R}$. By partial integration we get

$$(\hat{f})'(y) \ = \ -2\pi i \int_{\mathbf{R}} xe^{-\pi x^2}e^{-2\pi i x y}dx = 2\pi i^2 y \int_{\mathbf{R}} e^{-2\pi i x y}e^{-\pi x^2}dx$$

$$= \ -2\pi y\hat{f}(y).$$

It follows that the derivative of $\hat{f}(y)e^{\pi y^2}$ vanishes, and $\hat{f}(y)e^{\pi y^2} = c$ is a constant. Its value is $c = \hat{f}(0) = \int_{\mathbf{R}} e^{-\pi x^2} dx = 1$, as is well known. ∎

THEOREM 5.13 Let Λ be a lattice in $\mathbf{R}^n$. Then

$$\Theta\left(-\frac{1}{z}, \Lambda\right) = \left(\sqrt{-iz}\right)^n \frac{1}{\mathrm{vol}(\Lambda)} \cdot \Theta(z, \Lambda^*),$$

where the square root is positive for z on the positive imaginary axis.

Proof. Both sides in the asserted identity are holomorphic functions on the upper half-plane $\mathcal{H}$. Therefore it suffices to prove the identity for $z = it$, $t > 0$. We apply Theorem 5.11 to the function $f(x) = e^{-\pi x^2/t}$ on $\mathbf{R}^n$ whose Fourier transform is $\hat{f}(y) = \sqrt{t}^n e^{-\pi t y^2}$, by Lemma 5.12. Clearly, f satisfies the conditions (1), (2), (3) in Theorem 5.11. So we get

$$\begin{aligned}
\Theta\left(-\frac{1}{it}, \Lambda\right) &= \textstyle\sum_{v \in \Lambda} e^{-\pi v^2/t} = \frac{1}{\mathrm{vol}(\Lambda)} \sum_{u \in \Lambda^*} \hat{f}(u) \\
&= \frac{1}{\mathrm{vol}(\Lambda)} \textstyle\sum_{u \in \Lambda^*} \sqrt{t}^n e^{-\pi t u^2} \\
&= \sqrt{t}^n \frac{1}{\mathrm{vol}(\Lambda)} \Theta(it, \Lambda^*). \quad∎
\end{aligned}$$

Example 5.4, continued. The theta function of the one-dimensional self-dual lattice $\mathbf{Z}$ is Jacobi's theta function $\Theta(z, \mathbf{Z}) = \theta(z)$. In this case, Theorem 5.13 yields $\theta(-1/z) = \sqrt{-iz}\,\theta(z)$. The relation $\theta(z + 2) = \theta(z)$ is clear from the Fourier expansion of $\theta(z)$. This shows that $\theta(z)$ is a modular form of weight $1/2$ on the theta group Γ_θ.

If n is even then of course we need not care about the sign of the square root in $(\sqrt{-iz})^n = (-iz)^{n/2}$ in Theorem 5.13. We apply this theorem to even unimodular lattices and use the well-known fact that the full modular group Γ is generated by the transformations $T(z) = z + 1$ and $S(z) = -1/z$.

THEOREM 5.14 If Λ is an even unimodular lattice in $\mathbf{R}^n$, then n is a multiple of 8, and $\Theta(z, \Lambda)$ is a modular form of weight $n/2$ on the full modular group.

Proof. Since $\Lambda^* = \Lambda$ and $\mathrm{vol}(\Lambda) = 1$, Theorem 5.13 implies that

$$\Theta(S(z), \Lambda) = \Theta\left(-\frac{1}{z}, \Lambda\right) = (\sqrt{-iz})^n \Theta(z, \Lambda).$$

Since Λ is even, we have $\Theta(T(z), \Lambda) = \Theta(z+1, \Lambda) = \Theta(z, \Lambda)$. It follows that

$$\Theta((TS)(z), \Lambda) = (\sqrt{-iz})^n \Theta(z.\Lambda).$$

We apply this formula three times and observe that $(TS)^3 = -I$ is the identity on $\mathcal{H}$. This gives

$$\begin{aligned}
\Theta(z, \Lambda) &= \Theta((TS)^3(z), \Lambda) \\
&= \left(\sqrt{\frac{i}{z-1}}\right)^n \left(\sqrt{-i\frac{z-1}{z}}\right)^n (\sqrt{-iz})^n \Theta(z, \Lambda).
\end{aligned}$$

We put $z = i$. Since $\Theta(i, \Lambda)$ is a positive number, we obtain

$$\left(\sqrt{\frac{i}{i-1}}\right)^n \left(\sqrt{-(i-1)}\right)^n = 1.$$

Now from $i/(i-1) = (1/\sqrt{2})e^{-2\pi i/8}$ and $-(i-1) = \sqrt{2}e^{-2\pi i/8}$ we get

$$e^{-2\pi in/8} = 1.$$

So we conclude that n is a multiple of 8. Now the transformation formulas read

$$\Theta\left(-\frac{1}{z}, \Lambda\right) = z^{n/4}\Theta(z, \Lambda), \qquad \Theta(z+1, \Lambda) = \Theta(z, \Lambda).$$

Since Γ is generated by S and T, $\Theta(z, \Lambda)$ transforms like a modular form of weight $n/2$ on Γ. The Fourier expansion shows that it is holomorphic at the cusp ∞. So $\Theta(z, \Lambda)$ is a modular form of weight $n/2$ on Γ. ∎

COROLLARY 5.15 If $C \subseteq \mathbf{F}_2{}^n$ is a doubly-even self-dual code, then n is a multiple of 8.

Proof. By Proposition 5.4, the associated lattice Λ_C in $\mathbf{R}^n$ is even and unimodular. Therefore the assertion follows from Theorem 5.14. ∎

REMARK 5.1 For every positive integer $n \equiv 0 \pmod 8$, even unimodular lattices in $\mathbf{R}^n$ exist. Examples are given by the lattices Γ_n which were constructed in Example 5.3, or by the orthogonal direct sum of $n/8$ copies of the lattice E_8. We want to identify the theta series of the lattices E_8, Γ_8, $E_8 \oplus E_8$, and Γ_{16} with well-known modular forms.

The dimensions of the spaces $M_k(\Gamma) = M_k(\Gamma, 1)$ and $S_k(\Gamma) = S_k(\Gamma, 1)$ of modular forms respectively of cusp forms of weight k on the full modular group Γ are computed, as is well-known, by counting the number of zeroes of modular forms by a contour integration. The dimensions are 0 if k is odd or $k < 0$ or $k = 2$, and we have $S_0(\Gamma) = \{0\}$, $M_0(\Gamma) = \mathbf{C}$, and for even $k \geq 4$ the dimensions are

$$\mathrm{dim} M_k(\Gamma) \;=\; 1 + \mathrm{dim} S_k(\Gamma)$$

$$= \begin{cases} \lfloor k/12 \rfloor & \text{if } k \equiv 2 \pmod{12}, \\ 1 + \lfloor k/12 \rfloor & \text{if } k \not\equiv 2 \pmod{12}, \end{cases}$$

where $\lfloor x \rfloor$ denotes the greatest integer $\leq x$ for $x \in \mathbf{R}$. In particular, we have $S_k(\Gamma) = \{0\}$ and $\mathrm{dim} M_k(\Gamma) = 1$ for $k \in \{4, 6, 8, 10, 14\}$. These one-dimensional spaces are spanned by the *normalized Eisenstein series* $E_k(z)$ of weight k which can be defined, for any even $k \geq 2$, by the formula

$$E_k(z) = 1 - \frac{2k}{B_k} \sum_{m=1}^{\infty} \sigma_{k-1}(m) q^m \;;$$

here $\sigma_{k-1}(m) = \sum_{d\mid m} d^{k-1}$, and the *Bernoulli numbers* $B_2 = 1/6$, $B_4 = -1/30$, $B_6 = 1/42$, $B_8 = -1/30$, ... are defined by the power series expansion $z/(e^z - 1) = \sum_{k=0}^{\infty} (B_k/k!)z^k$ for $|z| < 2\pi$. There is a clash of notation with the E_8-lattice and the normalized Eisenstein series of weight 8. But tradition commands that we stick to the same symbol E_8 for these two different objects.

EXAMPLE 5.5 By Theorem 5.14, the theta series $\Theta(z, E_8)$ and $\Theta(z, \Gamma_8)$ both belong to $M_4(\Gamma)$. They both have the constant coefficient 1 in their Fourier expansion. Since $M_4(\Gamma)$ is spanned by the normalized Eisenstein series $E_4(z)$, it follows that

$$\Theta(z, E_8) = \Theta(z, \Gamma_8) = E_4(z) = 1 + 240 \sum_{m=1}^{\infty} \sigma_3(m) q^m.$$

This tells us that the number of vectors v in E_8 or in Γ_8 with $v^2 = 2m$ is equal to $240\sigma_3(m)$ for every integer $m \geq 1$. In the same way, since $M_8(\Gamma)$ is spanned by the normalized Eisenstein series $E_8(z)$, we obtain

$$\Theta(z, E_8 \oplus E_8) = \Theta(z, \Gamma_{16}) = E_8(z) = 1 + 480 \sum_{m=1}^{\infty} \sigma_7(m) q^m.$$

Thus the number of vectors v in $E_8 \oplus E_8$ or in Γ_{16} with $v^2 = 2m$ is equal to $480\sigma_7(m)$ for every integer $m \geq 1$. In Example 5.3 we have already shown that the lattices $E_8 \oplus E_8$ and Γ_{16} are not isomorphic.

REMARK 5.2 It is well-known that every modular form on Γ can uniquely be written as a polynomial in the normalized Eisenstein series $E_4(z)$ and $E_6(z)$. Thus the dimension formulae in the previous remark yield the identities $E_8(z) = E_4(z)^2$, $E_{10}(z) = E_4(z)E_6(z)$, $E_{14}(z) = E_4(z)^2 E_6(z)$. The smallest weight for which a non-trivial cusp form on Γ exists is $k = 12$. The normalized cusp form in $S_{12}(\Gamma)$ is denoted by $\Delta(z)$; it is called the *discriminant* in the theory of elliptic modular forms. It is related to the discriminant of elliptic curves over the field of complex numbers via the parametrization of these curves by the Weierstrass elliptic functions $\wp$ and $\wp'$. Since $E_4(z)^3 - E_6(z)^2$ belongs to $S_{12}(\Gamma)$, the computation of the first Fourier coefficient of this function shows that

$$\Delta(z) = 12^{-3}(E_4(z)^3 - E_6(z)^2).$$

From the definition of $E_4(z)$ and $E_6(z)$ it is easy to deduce that the coefficients in

$$\Delta(z) = \sum_{m=1}^{\infty} \tau(m) q^m$$

are integers. They are called the *Ramanujan numbers*. On the other hand, from Theorem 3.2 and $\dim S_{12}(\Gamma) = 1$ it follows that $\Delta(z)$ is a

Hecke eigenform. Therefore, the sequence of numbers $\tau(m)$ is multiplicative, it satisfies the recursion formula

$$\tau(p^{r+1}) = \tau(p)\tau(p^r) - p^{11}\tau(p^{r-1})$$

for powers of a prime p, and the corresponding Dirichlet series has an Euler product expansion

$$L(s, \Delta) = \prod_p \left(1 - \tau(p)p^{-s} + p^{11-2s}\right)^{-1}.$$

Ramanujan's empirical discovery of these facts in 1916 gave the starting point of what became the theory of Hecke operators.

2.2   The MacWilliams identity

Let $C \subseteq \mathbf{F}_2{}^n$ be a binary linear code and $\Lambda_C = (1/\sqrt{2})\rho^{-1}(C)$ the associated lattice in $\mathbf{R}^n$. The theta function of Λ_C is related to the weight enumerator of C in the following remarkable way.

THEOREM 5.16 Let Λ_C be the lattice associated to a binary linear code $C \subseteq \mathbf{F}_2{}^n$. Then the theta function of Λ_C and the weight enumerator of C satisfy the identity

$$\Theta(z, \Lambda_C) = W_C(A(z), B(z)),$$

where $A(z) = \theta(2z) = \Theta(z, \mathbf{Z})$ and

$$B(z) = \theta(\tfrac{z}{2}) - \theta(2z) = \sum_{m=-\infty}^{\infty} q^{(2m+1)^2/4}.$$

Proof. We use the notations from the proof of Proposition 5.4 and write vectors $v \in \Lambda_C$ uniquely in the form $v = (1/\sqrt{2})(\dot{c} + 2x)$ with $c \in C$ and $x \in \mathbf{Z}^n$. For a fixed codeword $c \in C$ we get

$$\sum_{v \in \rho^{-1}(c)} q^{\frac{1}{2}v^2} = \sum_{x \in \mathbf{Z}^n} q^{\frac{1}{4}(\dot{c}+2x)^2} = \sum_{x \in \mathbf{Z}^n} \prod_{j=1}^{n} q^{\frac{1}{4}(\dot{c}_j + 2x_j)^2}$$
$$= \prod_{j=1}^{n} \sum_{m=-\infty}^{\infty} q^{\frac{1}{4}(\dot{c}_j + 2m)^2}.$$

The series on m is equal to $\theta(2z) = A(z)$ if $\dot{c}_j = 0$, and equal to $B(z)$ if $\dot{c}_j = 1$. Therefore the product is equal to $A(z)^{n-w(c)}B(z)^{w(c)}$, where $w(c)$ is the weight of the codeword c. The summation over all codewords gives

$$\Theta(z, \Lambda_C) = \sum_{v \in \Lambda_C} q^{\frac{1}{2}v^2} = \sum_{c \in C} A(z)^{n-w(c)}B(z)^{w(c)} = W_C(A(z), B(z)). \quad \blacksquare$$

EXAMPLE 5.6 The lattice E_8 is associated to the extended Hamming code $\tilde{H}$ with weight distribution $W_{\tilde{H}} = X^8 + 14X^4Y^4 + Y^8$. From Example 5.5 we know that the theta function of E_8 is the normalized Eisenstein series $E_4(z)$ of weight 4. So from Theorem 5.16 we obtain the identity

$$E_4(z) = \theta(2z)^8 + 14\theta(2z)^4 \left(\theta(z/2) - \theta(2z)\right)^4 + \left(\theta(z/2) - \theta(2z)\right)^8.$$

Thus $E_4(z)$ is a polynomial in $\theta(2z)$ and $\theta(z/2)$.

We can also identify the discriminant function $\Delta(z)$ with a polynomial in $\theta(2z)$ and $\theta(z/2)$. From $\theta(-1/z) = \sqrt{-iz}\,\theta(z)$ and the definitions $A(z) = \theta(2z)$, $B(z) = \theta(z/2) - \theta(2z)$ it is easy to deduce the transformation formulas

$$A\left(-\tfrac{1}{z}\right) = \sqrt{-iz}\,\tfrac{1}{\sqrt{2}}\left(A(z) + B(z)\right),$$

$$B\left(-\tfrac{1}{z}\right) = \sqrt{-iz}\,\tfrac{1}{\sqrt{2}}\left(A(z) - B(z)\right).$$

In matrix notation they are rewritten as

$$\begin{pmatrix} A(-1/z) \\ B(-1/z) \end{pmatrix} = \sqrt{-iz}\,\frac{1}{\sqrt{2}}\begin{pmatrix} 1 & 1 \\ 1 & -1 \end{pmatrix} \cdot \begin{pmatrix} A(z) \\ B(z) \end{pmatrix}.$$

It is easily seen that the polynomial

$$P = A^4 B^4 (A^4 - B^4)^4$$

of degree 24 is invariant with respect to the orthogonal matrix $\frac{1}{\sqrt{2}}\begin{pmatrix} 1 & 1 \\ 1 & -1 \end{pmatrix}$ operating on the vector $\begin{pmatrix} A \\ B \end{pmatrix}$. (Observe that the factors A^2, B^2, $A^2 - B^2$, $A^2 + B^2$ are transformed into $(1/2)(A+B)^2$,

$(1/2)(A - B)^2$, $2AB$, $A^2 + B^2$, respectively.) This shows that

$$P\left(-\frac{1}{z}\right) = (\sqrt{-iz})^{24} P(z) = z^{12} P(z).$$

Clearly we have $A(z + 1) = A(z)$, and the definition of $B(z) = \sum_{m \text{ odd}} q^{m^2/2}$ yields $B(z + 1) = iB(z)$. It follows that $P(z + 1) = P(z)$. At the cusp ∞, $A(z)$ is holomorphic and $B(z)$ vanishes. This shows that $P(z)$ is a cusp form of weight 12 on the full modular group Γ. Since $S_{12}(\Gamma)$ is spanned by the discriminant $\Delta(z)$, we obtain $P(z) = c_1 \Delta(z)$, where the constant c_1 is the first Fourier coefficient of $P(z)$. Since $A(z) = 1 + 2q + \ldots$ and $B(z) = 2q^{1/4} + \ldots$, we have $c_1 = 16$. So we obtain the identity

$$\theta(2z)^4 \left(\theta(z/2) - \theta(2z)\right)^4 \left(\theta(2z)^4 - (\theta(z/2) - \theta(2z))^4\right)^4$$
$$= 16 \, \Delta(z) = \frac{1}{108} (E_4(z)^3 - E_6(z)^2).$$

THEOREM 5.17 (THE MACWILLIAMS IDENTITY) Let C be a binary $[n, k]$-code. Then the weight enumerator of C and that of the dual code $C^\perp$ are related by the identity

$$W_{C^\perp}(X, Y) = 2^{-k} W_C(X + Y, X - Y).$$

Proof. We use the transformation formula in Theorem 5.13 for $\Lambda = \Lambda_C$. We know that $\text{vol}(\Lambda_C) = 2^{\frac{n}{2} - k}$ from the beginning of Section 5.1.2. Proposition 5.9 tells us that $\Lambda_C^* = \Lambda_{C^\perp}$. Then Theorem 5.16 gives us

$$
\begin{aligned}
W_C\left(A(-1/z), B(-1/z)\right) &= \Theta\left(-\tfrac{1}{z}, \Lambda_C\right) \\
&= (\sqrt{-iz})^n \frac{1}{\text{vol}(\Lambda_C)} \Theta(z, \Theta_C^*) \\
&= (\sqrt{-iz})^n \, 2^{k - \frac{n}{2}} \, \Theta(z, \Lambda_{C^\perp}) \\
&= (\sqrt{-iz})^n \, 2^{k - \frac{n}{2}} \, W_{C^\perp}(A(z), B(z)).
\end{aligned}
$$

The transformation formulas for $A(-1/z)$ and $B(-1/z)$ in Example 5.6 yield, since W_C is a homogeneous polynomial of degree n, that

$$W_C(A(-1/z), B(-1/z))$$

$$= W_C\left(\sqrt{-iz/2}\,(A(z) + B(z)), \sqrt{-iz/2}\,(A(z) - B(z))\right)$$

$$= (\sqrt{-iz})^n\, 2^{-\frac{n}{2}}\, W_C(A(z) + B(z), A(z) - B(z)).$$

So we obtain the identity

$$W_{C^\perp}(A(z), B(z)) = 2^{-k}W_C(A(z) + B(z), A(z) - B(z)).$$

This will prove the MacWilliams identity once we know that the functions $A(z)$ and $B(z)$ are algebraically independent. Since $W_C(X, Y)$ is a homogeneous polynomial, it suffices to show that $A(z)$ and $B(z)$ don't annihilate a non-zero homogeneous polynomial. To the contrary, we assume that there were a homogeneous polynomial $F \neq 0$ in two variables such that $F(A(z), B(z))$ is identically 0 on $\mathcal{H}$. Then $f(Z) = F(1, Z)$ is a non-zero polynomial in one variable such that $f(B(z)/A(z))$ vanishes identically on $\mathcal{H}$. This means that the function $B(z)/A(z)$ can take as values only the finitely many roots of f. Since this function is meromorphic, it would be a constant. The constant would be 0, since the function vanishes at the cusp ∞. So $B(z) = \sum_{m=-\infty}^{\infty} q^{(2m+1)^2/4}$ would be identically 0, which is absurd. ∎

REMARK 5.3 We consider the weight enumerator in the inhomogeneous form

$$W_C(Y) = W_C(1, Y)$$

as in Section 3.3.1. From Theorem 5.17 we deduce

$$W_{C^\perp}(1, Y) = 2^{-k}W_C(1 + Y, 1 - Y) = 2^{-k}(1 + Y)^n W_C\left(1, \frac{1 - Y}{1 + Y}\right).$$

So we get the MacWilliams identity

$$W_{C^\perp}(Y) = 2^{-k}(1 + Y)^n W_C\left(\frac{1 - Y}{1 + Y}\right)$$

which is easily transformed into the form given in Theorem 3.5.

3. Doubly-even self-dual codes

3.1 The weight enumerator of doubly-even self-dual codes

In the special case of a self-dual code the MacWilliams identity reads as follows.

COROLLARY 5.18 Let $C \subseteq \mathbf{F}_2{}^n$ be a self-dual code. Then the weight enumerator $W_C(X, Y)$ of C is invariant with respect to the orthogonal transformation $\dfrac{1}{\sqrt{2}} \begin{pmatrix} 1 & 1 \\ 1 & -1 \end{pmatrix}$; it satisfies the identity

$$W_C(X, Y) = W_C\left(\frac{1}{\sqrt{2}}(X + Y), \frac{1}{\sqrt{2}}(X - Y)\right).$$

Now we can show that the weight distribution of a doubly-even self-dual code has a rather special form:

LEMMA 5.19 Every modular form f on the full modular group Γ whose weight k is a multiple of 4 can be written as a polynomial in the Eisenstein series E_4 and the discriminant function Δ.

THEOREM 5.20 (GLEASON'S THEOREM) Let $C \subseteq \mathbf{F}_2{}^n$ be a doubly-even self-dual code. Then the weight enumerator $W_C(X, Y)$ of C can be written as a polynomial in $Q(X, Y)$ and $P(X, Y)$, where

$$Q(X, Y) = W_{\tilde{H}}(X, Y) = X^8 + 14X^4Y^4 + Y^8,$$

$$P(X, Y) = X^4Y^4(X^4 - Y^4)^4.$$

Proofs. We mentioned the well-known fact that every $f \in M_k(\Gamma)$ can uniquely be written as a polynomial in the Eisenstein series E_4 and E_6. In this representation, every term has the weight k. Therefore, if $k \equiv 0 \pmod 4$, it follows that f is a polynomial in E_4 and E_6^2. Since $E_6^2 = E_4^3 - 12^3\Delta$, we obtain the assertion of the lemma.

Now let $C \subseteq \mathbf{F}_2{}^n$ be a doubly-even self-dual code. From Corollary 5.15 we know that n is a multiple of 8. From Theorem 5.16, Proposition 5.4 and Theorem 5.14 it follows that $W_C(A(z), B(z))$ is a modular form on Γ of weight $n/2 \equiv 0 \pmod 4$. So by Lemma 5.19 it is a polynomial in $E_4(z)$ and $\Delta(z)$. From Example 5.6 we know that $E_4(z) = W_{\tilde{H}}(A(z), B(z)) = Q(A(z), B(z))$ and $\Delta(z) = (1/16) P(A(z), B(z))$. This proves Theorem 5.20, since we have already shown that $A(z)$ and $B(z)$ are algebraically independent. ∎

EXAMPLE 5.7 We suppose that $C \subseteq \mathbf{F}_2{}^{24}$ is a doubly-even self-dual code of length 24 with minimal distance $d \geq 8$. For the associated lattice Λ_C this means that its only roots are the 48 vectors $\pm\sqrt{2}e_\nu$ which come from preimages of the codeword 0. From Theorem 5.20 it follows that

$$W_C(X, Y) = Q(X, Y)^3 + aP(X, Y) = X^{24} + (42 + a)X^{20}Y^4 + \cdots .$$

The condition $d \geq 8$ implies that $42 + a = 0$. Therefore we get

$$
\begin{aligned}
W_C(X, Y) &= Q(X, Y)^3 - 42P(X, Y) \\
&= X^{24} + 759X^{16}Y^8 + 2576X^{12}Y^{12} + 759X^8Y^{16} + Y^{24}.
\end{aligned}
$$

As a result we see that the weight distribution of such a code is uniquely determined and, in particular, that there are exactly 759 codewords of minimal weight 8.

In the same way, if $C \subseteq \mathbf{F}_2{}^{32}$ is a doubly-even self-dual code of length 32 with minimal distance $d \geq 8$, it follows that the weight enumerator of C is

$$
\begin{aligned}
W_C(X, Y) &= Q(X, Y)^4 - 56\, Q(X, Y)P(X, Y) \\
&= X^{32} + 620X^{24}Y^8 + 13888X^{20}Y^{12} + 36518X^{16}Y^{16} \\
&\qquad + 13888X^{12}Y^{20} + 620X^8Y^{24} + Y^{32}.
\end{aligned}
$$

Of course, these results don't tell us anything about the existence of such codes.

More generally, for any $n > 0$ which is a multiple of 8 we ask for a doubly-even self-dual code $C \subseteq \mathbf{F}_2{}^n$ whose minimal distance d is as large as possible. By Theorem 5.20, the weight enumerator $W_C(X, Y) =$

$\sum_{i=0}^{n} A_i X^{n-i} Y^i$ is a polynomial in $Q(X, Y) = X^8 + 14X^4Y^4 + Y^8$ and $P(X, Y) = X^4 Y^4 (X^4 - Y^4)^4$. We write

$$n = 24m + 8l \qquad \text{with} \qquad l \in \{0, 1, 2\}.$$

Then the weight enumerator can be written as

$$W_C = Q^{3m+l} + \sum_{j=1}^{m} \alpha_j P^j Q^{3(m-j)+l} = \sum_{j=0}^{m} \alpha_j P^j Q^{3(m-j)+l}$$

with integers $\alpha_0 = 1, \alpha_1, \ldots, \alpha_m$. Our condition on d means that in W_C as many consecutive coefficients $A_4, A_8, \ldots$ as possible vanish. Any equation $A_{4\nu} = 0$ is equivalent to an inhomogeneous linear equation for $\alpha_1, \ldots, \alpha_m$. Therefore it is to be expected that one can achieve that $A_4 = \ldots = A_{4m} = 0$ and that the further coefficients $A_{4m+4}, \ldots$ are uniquely determined. So we have $d \geq 4m + 4$.

A doubly-even self-dual code $C \subseteq \mathbf{F}_2^n$ of length $n = 24m + l$ is called an *extremal code* if its minimal distance d satisfies $d \geq 4m+4 = 4\lfloor n/24 \rfloor + 4$.

It can be shown that indeed $d = 4m+4$. Also, it can be shown that extremal codes can exist only for finitely many lengths n. Examples of extremal codes are known for $n = 8, 16, 24, 32, 40, 48, 56, 64, 80, 88, 104, 136$. For more details, see [4], pp. 193 - 195. Clearly, the extended Hamming code $\tilde{H}$ and the direct sum $\tilde{H} \oplus \tilde{H}$ are extremal codes of lengths 8 and 16, respectively. In Section 5.3.2 we will construct the extended Golay code $\tilde{G}$ which is an extremal code of length 24.

There are analogous problems and results for even unimodular lattices Λ in $\mathbf{R}^n$. As before we write $n = 24m + 8l$ with $l \in \{0, 1, 2\}$. We ask for a lattice Λ whose shortest vectors $v \neq 0$ are as long as possible. This means that in the theta function

$$\Theta(z, \Lambda) = \sum_{\nu=0}^{\infty} a_\nu q^\nu$$

as many consecutive coefficients $a_1, a_2, a_3, \ldots$ as possible vanish. The proof of Theorem 5.20 has shown that $\Theta(z, \Lambda)$ is a polynomial in $E_4(z)$ and $\Delta(z)$. So we have

$$\Theta(z,\Lambda) = E_4(z)^{3m+l} + \sum_{j=1}^{m} \alpha_j \Delta(z)^j E_4(z)^{3(m-j)+l}$$

with rational numbers $\alpha_1, \ldots, \alpha_m$. Any equation $a_\nu = 0$ is equivalent to an inhomogeneous linear equation for $\alpha_1, \ldots, \alpha_m$. Therefore it is to be expected that one can achieve $a_1 = \ldots = a_m = 0$ and that the further coefficients $a_{m+1}, a_{m+2}, \ldots$ are uniquely determined. An even unimodular lattice Λ in $\mathbf{R}^{24m+8l}$ is called an *extremal lattice* if the Fourier expansion of its theta function starts with

$$\Theta(z,\Lambda) = 1 + a_{m+1} q^{m+1} + \ldots .$$

Similarly to the case of extremal codes, it can be shown that extremal lattices can exist only in finitely many dimensions. The shortest vectors $v \neq 0$ in an extremal lattice $\Lambda \in \mathbf{R}^{24m+8l}$ have a squared length $v^2 \geq 2m + 2$. We observe that for $m \geq 1$ the lattice Λ_C associated to an extremal code $C \subseteq \mathbf{F}_2^{24m+8l}$ cannot be an extremal lattice, since in Λ_C there are $2n = 48m + 16l$ roots which come from preimages of the codeword 0. Nevertheless it is possible to construct the famous Leech lattice, which is an extremal lattice in $\mathbf{R}^{24}$, with the help of the extended Golay code $\tilde{\mathcal{G}}$, an extremal code in $\mathbf{F}_2^{24}$. This will be done in Section 5.3.3.

3.2   The extended Golay code

The extended Golay code $\tilde{\mathcal{G}}$ was invented by M. J. E. Golay in 1949. It is an extremal code in $\mathbf{F}_2^{24}$. So its minimal distance is $d = 8$, it can correct 3 errors, and the information rate is $1/2$. Several constructions are known for this code. We choose a construction which starts from the geometry of an icosahedron.

DEFINITION 5.10 We consider the icosahedron as a graph. This graph has 12 vertices $V_1, \ldots, V_{12}$. Every vertex is connected by an edge with exactly 5 neighbouring vertices, as shown in the diagram.

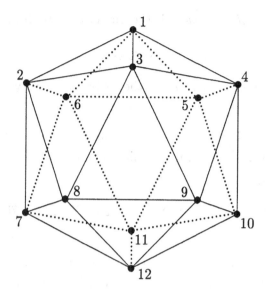

Now we define the 12 by 12 matrix $A = (a_{ij})$ to be the complement of the adjacency matrix of the icosahedron, i.e., the entries a_{ij} are given by $a_{ij} = 0$ if the vertices V_i and V_j are connected by an edge, and $a_{ij} = 1$ otherwise. We obtain the symmetric matrix

$$A = \begin{pmatrix}
1 & 0 & 0 & 0 & 0 & 0 & 1 & 1 & 1 & 1 & 1 & 1 \\
0 & 1 & 0 & 1 & 1 & 0 & 0 & 0 & 1 & 1 & 1 & 1 \\
0 & 0 & 1 & 0 & 1 & 1 & 1 & 0 & 0 & 1 & 1 & 1 \\
0 & 1 & 0 & 1 & 0 & 1 & 1 & 1 & 0 & 0 & 1 & 1 \\
0 & 1 & 1 & 0 & 1 & 0 & 1 & 1 & 1 & 0 & 0 & 1 \\
0 & 0 & 1 & 1 & 0 & 1 & 0 & 1 & 1 & 1 & 0 & 1 \\
1 & 0 & 1 & 1 & 1 & 0 & 1 & 0 & 1 & 1 & 0 & 0 \\
1 & 0 & 0 & 1 & 1 & 1 & 0 & 1 & 0 & 1 & 1 & 0 \\
1 & 1 & 0 & 0 & 1 & 1 & 1 & 0 & 1 & 0 & 1 & 0 \\
1 & 1 & 1 & 0 & 0 & 1 & 1 & 1 & 0 & 1 & 0 & 0 \\
1 & 1 & 1 & 1 & 0 & 0 & 0 & 1 & 1 & 0 & 1 & 0 \\
1 & 1 & 1 & 1 & 1 & 1 & 0 & 0 & 0 & 0 & 0 & 1
\end{pmatrix}.$$

The *extended Golay code* $\tilde{\mathcal{G}}$ is the binary code with the generator matrix

$$G = (I_{12} \quad A)$$

in standard form, where I_{12} is the 12 by 12 identity matrix.

PROPOSITION 5.21 The extended Golay code $\tilde{\mathcal{G}}$ is a doubly-even self-dual code in $\mathbf{F}_2{}^{24}$. Its minimal distance is $d = 8$. It is an extremal code, and its weight enumerator is

$$W_{\tilde{\mathcal{G}}}(X, Y) = X^{24} + 759X^{16}Y^8 + 2576X^{12}Y^{12} + 759X^8Y^{16} + Y^{24}.$$

Proof. The dimension of $\tilde{\mathcal{G}}$ is equal to the rank of G which is 12. Let $v_1, \ldots, v_{12}$ denote the rows of G. The weight of every row is $w(v_i) = 8$. For $i \neq j$, the rows v_i and v_j have exactly two entries 1 in common positions if they correspond to antipodal vertices V_i and V_j of the icosahedron, and otherwise they have exactly four entries 1 in common positions. It follows that $w(v_i + v_j) = 12$ in the antipodal case and $w(v_i + v_j) = 8$ otherwise. In particular, we get $v_i \cdot v_j = 0$ in $\mathbf{F}_2$ for all i and j. Thus the code $\tilde{\mathcal{G}}$ is self-orthogonal, $\tilde{\mathcal{G}} \subseteq \tilde{\mathcal{G}}^\perp$, and because of $\dim(\tilde{\mathcal{G}}^\perp) = 12 = \dim(\tilde{\mathcal{G}})$ we get $\tilde{\mathcal{G}} = \tilde{\mathcal{G}}^\perp$; the code $\tilde{\mathcal{G}}$ is self-dual.

For any codeword $c \in \tilde{\mathcal{G}}$ we consider the unique preimage $\dot{c} \in \rho^{-1}(c)$ in $\mathbf{Z}^{24}$ all of whose coordinates are 0 or 1. Then we observe that $\dot{v}_i^2 = 8$ for all i and that $\dot{v}_i \cdot \dot{v}_j$ is even for $i \neq j$. Every codeword $c \in \tilde{\mathcal{G}}$ is a sum of rows of G. Therefore, $\dot{c}^2$ is a sum of certain terms $\dot{v}_i^2$ and $2\dot{v}_i \cdot \dot{v}_j$ which are multiples of 4. It follows that $w(c) = \dot{c}^2$ is a multiple of 4. Thus we have shown that the code $\tilde{\mathcal{G}}$ is doubly-even.

Since $\tilde{\mathcal{G}}$ is self-dual and

$$(I_{12} \quad A) \begin{pmatrix} A \\ I_{12} \end{pmatrix} = 0$$

over $\mathbf{F}_2$, the parity check matrix

$$\tilde{G} = (A \quad I_{12})$$

is also a generator matrix of $\tilde{\mathcal{G}}$.

Now we suppose that the codeword $c = v_i + v_j + v_k$ is the sum of three distinct rows of G. Then c has exactly 3 coordinates equal to 1 in the first block of 12 positions. We show that in the second block there are exactly 5 or 9 coordinates equal to 1, resulting in $w(c) = 8$ or 12: If none of the vertices V_i, V_j and V_k of the icosahedron are connected by an edge, then we may assume, because of the symmetry of the icosahedron, that $\{i, j, k\} = \{1, 7, 9\}$. Then we find $w(c) = 12$. If at least two of the vertices are connected by an edge, then we may assume that $\{i, j\} = \{1, 2\}$. In this case we find that $w(c) = 12$ if the subgraph of the three vertices is a triangle and $w(c) = 8$ otherwise. We have shown now that any codeword $c \in \tilde{\mathcal{G}}$ which is a sum of at most 3 rows of G satisfies $w(c) \geq 8$. Clearly, the same result holds if c is the sum of at most 3 rows of $\tilde{G}$.

Finally, we suppose that $c \in \tilde{\mathcal{G}}$ is a sum of l distinct rows of G where $l \geq 4$. Then c is also a sum of a certain number m of distinct rows of $\tilde{G}$. For the proof of $w(c) \geq 8$ it suffices to consider the case that $m \geq 4$. Then the word c has l coordinates equal to 1 in the first block of 12 positions, and it has m coordinates equal to 1 in the last block of 12 positions. It follows that $w(c) = l + m \geq 8$. Thus we have shown that the minimal distance of $\tilde{\mathcal{G}}$ is $d = 8$. Now it is clear from the definition that $\tilde{\mathcal{G}}$ is an extremal code. The formula for the weight enumerator of $\tilde{\mathcal{G}}$ follows from the discussion in Example 5.7 in Section 5.3.1. ∎

REMARK 5.4 It can be shown that any two doubly-even self-dual codes in $\mathbf{F}_2{}^{24}$ with minimal distance 8 are equivalent. Therefore the extended Golay code $\tilde{\mathcal{G}}$ is uniquely determined by its properties, up to equivalence, and it does not depend on a particular construction. We omit the proof of the uniqueness of $\tilde{\mathcal{G}}$.

The extended Hamming code $\tilde{H}$ was obtained from the Hamming code H by adding a parity check bit to every codeword of H. We get back H from $\tilde{H}$ by dropping the last coordinate of every codeword of $\tilde{H}$. In the same way, we can define the *Golay code* $\mathcal{G} \subseteq \mathbf{F}_2{}^{23}$ by dropping the last coordinate of every codeword of $\tilde{\mathcal{G}}$. Then $\mathcal{G}$ is a binary $[23, 12]$-code. Any row of G which has 1 as its last coordinate gives us a word in $\mathcal{G}$ with weight 7. Since $\tilde{\mathcal{G}}$ has minimal distance 8, it follows that the minimal distance of $\mathcal{G}$ is 7.

The Hamming code H and the Golay code $\mathcal{G}$ are examples of perfect codes:

DEFINITION 5.11 A code $C \subseteq \mathbf{F}_q^n$ with an odd minimal distance $d = 2e + 1$ is called a *perfect code* if for every $x \in \mathbf{F}_q^n$ there is exactly one codeword $c \in C$ such that the distance of x and c is at most e. Equivalently, $\mathbf{F}_q^n$ is the disjoint union of balls of radius e around the codewords of C. For $0 \leq \nu \leq e$ and any $c \in \mathbf{F}_q^n$ there are exactly $\binom{n}{\nu}(q-1)^\nu$ points $x \in \mathbf{F}_q^n$ such that $d(x,c) = \nu$. Therefore a code $C \subseteq \mathbf{F}_q^n$ with minimal distance $2e + 1$ is perfect if and only if

$$|C|\left(1 + \binom{n}{1}(q-1) + \binom{n}{2}(q-1)^2 + \ldots + \binom{n}{e}(q-1)^e\right) = q^n.$$

Now the relations

$$2^4\left(1 + \binom{7}{1}\right) = 2^7, \qquad 2^{12}\left(1 + \binom{23}{1} + \binom{23}{2} + \binom{23}{3}\right) = 2^{23}$$

prove that the Hamming code H and the Golay code $\mathcal{G}$ are perfect codes.

3.3   The Leech lattice

In this subsection the symbols Γ and $\tilde{\Gamma}$ stand for the lattice $\Gamma = \rho^{-1}(\tilde{\mathcal{G}}) \subseteq \mathbf{Z}^{24}$ and the lattice $\tilde{\Gamma} = \Lambda_{\tilde{\mathcal{G}}}$ which is associated to the extended Golay code $\tilde{\mathcal{G}}$. From Propositions 5.21 and 5.4 it follows that $\tilde{\Gamma}$ is an even unimodular lattice. We know that there are exactly 48 roots in $\tilde{\Gamma}$. Therefore, in the theta function $\Theta(z, \tilde{\Gamma}) = \sum_{m=0}^\infty a_m q^m$ we have $a_0 = 1$ and $a_1 = 48$. Since $\Theta(z, \tilde{\Gamma})$ is a modular form of weight 12 on the full modular group and since the space of these modular forms has dimension 2, it follows that $\Theta(z, \tilde{\Gamma})$ is uniquely determined by the data $a_0 = 1$, $a_1 = 48$. In particular, we can compute the number a_2 of vectors $v \in \tilde{\Gamma}$ with squared length $v^2 = 4$:

LEMMA 5.22 Let $f(z) = \sum_{m=0}^{\infty} a_m q^m$ be a modular form of weight 12 on the full modular group. Then we have

$$a_2 = 196560\,a_0 - 24\,a_1.$$

In the lattice $\tilde{\Gamma}$ which is associated to the extended Golay code $\tilde{\mathcal{G}}$, there are exactly $196560 - 24 \cdot 48 = 195408$ vectors v such that $v^2 = 4$.

Proof. It suffices to prove the assertion for the functions E_6^2 and Δ which form a basis of the space of modular forms of weight 12. A straightforward calculation shows that

$$E_6(z)^2 = 1 - 1008\,q + 220752\,q^2 + \ldots, \qquad \Delta(z) = q - 24\,q^2 + \ldots . \qquad \blacksquare$$

The coefficient a_2 in $\Theta(z, \tilde{\Gamma}) = \sum_{m=0}^{\infty} a_m q^m$ can also be computed directly from the weight distribution of the extended Golay code. One obtains

$$a_2 = \binom{24}{2} 2^2 + 759 \cdot 2^8 = 195408.$$

Now we define the Leech lattice Λ in $\mathbf{R}^{24}$ as follows.

DEFINITION 5.12 We use our previous notations and write $v \in \Gamma = \rho^{-1}(\tilde{\mathcal{G}})$ uniquely in the form $v = (v_1, \ldots, v_{24}) = \dot{c} + 2x$ with $c = (c_1, \ldots, c_{24}) \in \tilde{\mathcal{G}}$ and $x = (x_1, \ldots, x_{24}) \in \mathbf{Z}^{24}$. Since $\tilde{\mathcal{G}}$ is doubly-even, we have $\dot{c}_1 + \ldots + \dot{c}_{24} \in 4\mathbf{Z}$. Therefore, a map $\phi : \Gamma \to \mathbf{F}_2$ is defined by the formula

$$\phi(v) = \frac{1}{2}(v_1 + \ldots + v_{24}) \,(\mathrm{mod}\ 2) = x_1 + \ldots + x_{24} \,(\mathrm{mod}\ 2),$$

and ϕ is an epimorphism of additive groups. Thus the kernel

$$\Gamma_1 = \phi^{-1}(0)$$

is a sublattice of index 2 in Γ. The lattice Γ is the disjoint union of Γ_1 and the coset

$$\Gamma_2 = \phi^{-1}(1) = \{\dot{c} + 2x \ : \ c \in \tilde{\mathcal{G}},\ x \in \mathbf{Z}^{24},\ x_1 + \ldots + x_{24} \text{ is odd}\}.$$

Now we proceed similarly as in Example 5.3: Let $v_0 = (1/2)(e_1 + \ldots + e_{24}) = (1/2, \ldots, 1/2)$ where $e_1, \ldots, e_{24}$ are the standard unit vectors in $\mathbf{R}^{24}$. We consider the translate $v_0 + \Gamma_2$ and put

$$\Lambda = \frac{1}{\sqrt{2}} \left(\Gamma_1 \cup (v_0 + \Gamma_2) \right).$$

Since Γ_1 and Γ are lattices and $2v_0 \in \Gamma_1$, it follows that Λ is a lattice in $\mathbf{R}^{24}$. The lattice Λ is called the *Leech lattice*. It was discovered by J. Leech in 1965.

THEOREM 5.23 The Leech lattice Λ is an even unimodular lattice in $\mathbf{R}^{24}$. There are no roots in Λ. There are exactly 196560 vectors $v \in \Lambda$ such that $v^2 = 4$. The Leech lattice is an extremal lattice.

Proof.   Since $\tilde{\mathcal{G}}$ is a doubly-even self-dual code, the lattice $\tilde{\Gamma} = (1/\sqrt{2})\rho^{-1}(\tilde{\mathcal{G}})$ is even and unimodular. Therefore, $(1/\sqrt{2})\Gamma_1$ is an even lattice with index 2 in Λ. Any vector $v \in (1/\sqrt{2})(v_0 + \Gamma_2)$ can be written as $v = (1/\sqrt{2})(v_0 + \dot{c} + 2x)$ where $c \in \tilde{\mathcal{G}}$, $x \in \mathbf{Z}^{24}$ and $x_1 + \ldots + x_{24}$ is odd. It follows that

$$v^2 = \frac{1}{2}(v_0 + \dot{c} + 2x)^2 \equiv \frac{1}{2}v_0^2 + v_0 \cdot \dot{c} + 2v_0 \cdot x \equiv 3 + (x_1 + \ldots + x_{24}) \equiv 0 \ (\mathrm{mod}\ 2),$$

hence $v^2 \in 2\mathbf{Z}$. It follows that $u \cdot v = (1/\sqrt{2})((u+v)^2 - u^2 - v^2) \in \mathbf{Z}$ for all $u, v \in \Lambda$. Thus Λ is an even lattice.

The lattice $(1/\sqrt{2})\Gamma_1$ has index 2 in $\tilde{\Gamma}$ as well as in Λ. Therefore we get $\mathrm{disc}(\Lambda) = \mathrm{disc}(\tilde{\Gamma}) = 1$, and the lattice Λ is unimodular.

Now we show that there are no roots in Λ. Let $v = (1/\sqrt{2})(\dot{c} + 2x) \in (1/\sqrt{2})\Gamma_1$ and $v \neq 0$. If $c = 0$ then $v^2 = 2x^2 = 2(x_1^2 + \ldots + x_{24}^2) \geq 4$. If $c \neq 0$ then at least 8 coordinates in $\dot{c} + 2x$ are odd, and it follows that $v^2 \geq 4$ as well. For $v = (1/\sqrt{2})(v_0 + \dot{c} + 2x) \in (1/\sqrt{2})(v_0 + \Gamma_2)$, all coordinates of $v_0 + \dot{c} + 2x$ are halves of odd integers. This implies that $v^2 \geq (1/2) \cdot (1/4) \cdot 24 = 3$. Therefore there are no roots in $(1/\sqrt{2})(v_0 + \Gamma_2)$ and no roots in Λ. Now from Lemma 5.22 we get $a_2 = 196560$ for the number of vectors $v \in \Lambda$ satisfying $v^2 = 4$. ∎

REMARK 5.5 The *kissing number* in n-space is the maximal number of solid balls of equal size in $\mathbf{R}^n$ which can touch a further solid ball of the same size. This number is denoted by τ_n. (The term "kissing number" is borrowed from billiard language.) We consider balls of radius 1 in $\mathbf{R}^{24}$ with centers at each point of the Leech lattice. This configuration of balls shows that $\tau_{24} \geq 196560$. In the same way, the properties of the lattice E_8 show that $\tau_8 \geq 240$. In fact it can be proved (see [4]) that

$$\tau_{24} = 196560, \qquad \tau_8 = 240,$$

and curiously these are the only known kissing numbers except for the low-dimensional cases $\tau_3 = 12$, $\tau_2 = 6$, and the trivial case $\tau_1 = 2$.

Appendix A

The Kloosterman Codes and Distribution of the Weights[1]

1.   Introduction

Let p be a prime, and let $q = p^r$ for some positive integer $r \geq 2$. We denote the finite field with q elements by $\mathbf{F}_q$, and we fix a primitive element of $\mathbf{F}_q$ which we denote by α (i.e. $\mathbf{F}_q^\times = \langle \alpha \rangle$). We also define $e(x) = e^{2\pi i x / p}$ and the trace function tr by

$$\operatorname{tr} \beta := \beta + \beta^p + \cdots + \beta^{p^{r-1}} \quad (\in \mathbf{F}_p) \tag{A.1}$$

for $\beta \in \mathbf{F}_q$. Our starting point is the Melas code, which is a classical cyclic code defined in 1960:

DEFINITION A.1 (MELAS [22], SEE ALSO SECTION 3.3.2) Let $m(x) \in \mathbf{F}_p[x]$ be the minimal polynomial of α and $m_-(x) \in \mathbf{F}_p[x]$ be that of α^{-1}. Then the cyclic code defined by the ideal

$$M(q) = (m(x)m_-(x)) \tag{A.2}$$

[1]The contents of this appendix are based on the article *Hyper-Kloosterman sums and coding theory* by K. Chinen and T. Hiramatsu in *Proc. Fourth Conference on Algebraic Geometry, Number Theory, Coding Theory and Cryptography, 21-24, November 2000, University of Tokyo*. They are essentially the same as the following two papers: Chinen-Hiramatsu [2] and K. Chinen; *On some properties of the hyper-Kloosterman codes*, to appear in Tokyo J. Math.

is called the *Melas code*.

The code $M(q)$ is a 2-error correcting code. Next we introduce the Kloosterman sum. It is a kind of exponential sums which we encounter so often in number theory:

DEFINITION A.2 For $a, b \in \mathbf{F}_q$, the sums

$$K(a, b; q) = \sum_{x \in \mathbf{F}_q^{\times}} e(\mathrm{tr}\,(ax + bx^{-1})) \qquad (A.3)$$

are called the *Kloosterman sums*.

The sums $K(a, b; q)$ were first introduced by Kloosterman [14] who dealt with a certain Diophantine equation, and they proved to be closely related to modular forms (see Kloosterman [15] and Petersson [26]).

These two notions, the Melas code and the Kloosterman sums, are related to each other by the fact that the Hamming weights of the codewords in $M(q)^{\perp}$ are expressed by the Kloosterman sums. For this property, $M(q)^{\perp}$ is sometimes called the Kloosterman code.

We sometimes meet in number theory and algebraic geometry, a generalization of (A.3), called the hyper-Kloosterman sums:

DEFINITION A.3 For any $a \in \mathbf{F}_q{}^m$, the *hyper-Kloosterman sums of degree* $m - 1$ are defined by

$$K_m(a; q) = \sum_{x \in (\mathbf{F}_q^{\times})^{m-1}} e(\mathrm{Tr}\,(a, x)), \qquad (A.4)$$

where

$$\mathrm{Tr}\,(a, x) = \mathrm{tr}\,(a_1 x_1 + a_2 x_2 + \cdots a_{m-1} x_{m-1} + a_m (x_1 x_2 \cdots x_{m-1})^{-1}) \quad (A.5)$$

for $a = (a_1, a_2, \cdots, a_m)$ and $x = (x_1, x_2, \cdots, x_{m-1})$.

REMARK A.1 Some authors call them $m - 1$ dimensional hyper-Kloosterman sums, but we use the word "degree" instead of "dimension" to avoid the confusion with the dimension of a code.

We can easily see that $K_2(a; q) = K(a, b; q)$ by letting $a = (a, b)$. The purpose of this article is to define a new code, which is a natural generalization of $M(q)^\perp$. It should be called the hyper-Kloosterman code, and we deduce various properties of it.

In Section A.2, we review the relation between $M(q)^\perp$ and $K(a, b; q)$. The Delsarte theorem for the subfield subcode and the trace code is used as a key lemma. In Section A.3, we introduce our new code, the hyper-Kloosterman code. The sections following it are devoted to stating the properties of the hyper-Kloosterman code: we discuss the quasi-cyclic property in Section A.4, weight distribution in Section A.5. Estimation of the minimum distance (when $p - 1|m$) is done in Section A.6, and in the last section, some divisibility theorem of the Hamming weight is deduced.

2. Melas code and Kloosterman sums

In this section we review the relation between $M(q)^\perp$ and $K(a, b; q)$. Essentially this is discussed in Lachaud [17], Wolfmann [37] and Hiramatsu [11], but repeating their results should not be meaningless because employment of the Delsarte theorem will make the argument clearer. Indeed the Delsarte theorem is not mentioned explicitly in any of the above papers. We begin with some reviews of Section 3.3.1:

DEFINITION A.4 Let $\check{C}$ be a code over $\mathbf{F}_q$. Then the code

$$C = \check{C} \cap \mathbf{F}_p{}^n \qquad (A.6)$$

is called the *subfield subcode* of $\check{C}$.

REMARK A.2 We can also define a subfield subcode of $\check{C}$ as a code over an intermediate field of $\mathbf{F}_q$ and $\mathbf{F}_p$, but for simplicity, we restrict ourselves to the subfield subcode over $\mathbf{F}_p$.

There is another class of codes, which are constructed over the subfields:

DEFINITION A.5 Let C be a code over $\mathbf{F}_p$. Then C is called a *trace code* if there exists a code $\bar{C}$ over $\mathbf{F}_q$ such that

$$C = \operatorname{tr} \bar{C} := \{(\operatorname{tr} c_1, \cdots, \operatorname{tr} c_n) \mid (c_1, \cdots, c_n) \in \bar{C}\}. \qquad (A.7)$$

The following theorem tells us the relation between the subfield subcode and the trace code, which is the same as Theorem 3.9 in Section 3.3.1:

THEOREM A.1 (DELSARTE) Let $\check{C}$ be a code over $\mathbf{F}_q$, and C be the subfield subcode of $\check{C}$. Then we have

$$C^\perp = \operatorname{tr}(\check{C}^\perp). \qquad (A.8)$$

Now we consider the Melas code and its dual. Recall that α is a primitive element of $\mathbf{F}_q$.

PROPOSITION A.2 Let $\check{M}(q)$ be a code over $\mathbf{F}_q$ with a parity check matrix

$$\check{H}_q = \begin{pmatrix} 1 & \alpha & \alpha^2 & \cdots & \alpha^{q-2} \\ 1 & \alpha^{-1} & \alpha^{-2} & \cdots & \alpha^{-(q-2)} \end{pmatrix}. \qquad (A.9)$$

Then the Melas code $M(q)$ is the subfield subcode of $\check{M}(q)$.

Proof. Take a vector $c = (c_0, c_1, \cdots, c_{q-2}) \in \mathbf{F}_p^{q-1}$. Then c is contained in the subfield subcode of $\check{M}(q)$ if and only if

$$c_0 + c_1\alpha + \cdots + c_{q-2}\alpha^{q-2} = c_0 + c_1\alpha^{-1} + \cdots + c_{q-2}\alpha^{-(q-2)} = 0,$$

which is equivalent to $c \in M(q)$. ∎

We obtain the following explicit representation of the dual code $\check{M}(q)^\perp$:

$$\check{M}(q)^\perp = \{c\check{H}_q \in \mathbf{F}_q{}^{q-1} \mid c \in \mathbf{F}_q{}^2\}$$
$$= \{(a + b, a\alpha + b\alpha^{-1}, \cdots, a\alpha^{q-2} + b\alpha^{-(q-2)}) \in \mathbf{F}_q{}^{q-1}$$
$$\mid (a, b) \in \mathbf{F}_q{}^2\} \quad (A.10)$$

which yields, via the Delsarte theorem and the proposition above, a representation of the dual of the Melas code:

$$M(q)^\perp = \text{tr}\,(\check{M}(q)^\perp)$$

$$= \{\varphi(a, b) := (\text{tr}\,(a + b), \text{tr}\,(a\alpha + b\alpha^{-1}), \cdots, \text{tr}\,(a\alpha^{q-2} + b\alpha^{-(q-2)}))$$
$$\in \mathbf{F}_p{}^{q-1} \mid (a, b) \in \mathbf{F}_q{}^2\}. \quad (A.11)$$

This expression leads to a formula for the Hamming weights of the code-words in $M(q)^\perp$:

THEOREM A.3 The Hamming weight of the codeword $\varphi(a, b) \in M(q)^\perp$ is given by

$$w(\varphi(a, b)) = \frac{1}{p}\Big\{(p - 1)(q - 1) - \sum_{s \in \mathbf{F}_p^\times} K(sa, sb; q)\Big\}. \quad (A.12)$$

Proof. Let $f(x) = ax + bx^{-1}$. For any $x \in \mathbf{F}_q$, we have

$$\sum_{s \in \mathbf{F}_p} e(s \cdot \text{tr}\, f(x)) = \begin{cases} p, & \text{if tr}\, f(x) = 0, \\ 0, & \text{if tr}\, f(x) \neq 0. \end{cases} \quad (A.13)$$

So the number of the entries of $\varphi(a, b)$ which are zero equals

$$\sharp\{x \in \mathbf{F}_q^\times \mid \text{tr}\, f(x) = 0\} = \sum_{x \in \mathbf{F}_q^\times} \frac{1}{p} \sum_{s \in \mathbf{F}_p} e(s \cdot \text{tr}\, f(x))$$

$$= \frac{1}{p} \sum_{s \in \mathbf{F}_p} \sum_{x \in \mathbf{F}_q^\times} e(\text{tr}\,(sf(x))). \quad (A.14)$$

Therefore the Hamming weight of $\varphi(a, b)$ is

$$w(\varphi(a, b)) = q - 1 - \frac{1}{p} \sum_{s \in \mathbf{F}_p} \sum_{x \in \mathbf{F}_q^\times} e(\text{tr}\,(sf(x)))$$

$$= q - 1 - \frac{1}{p}(q-1)$$

$$- \frac{1}{p} \sum_{s \in \mathbf{F}_p^\times} \sum_{x \in \mathbf{F}_q^\times} e(\operatorname{tr}(sf(x))). \qquad (A.15)$$

Noting that $\sum_{x \in \mathbf{F}_q^\times} e(\operatorname{tr}(sf(x))) = K(sa, sb, q)$, we get the theorem. ∎

Thus the relation between the Melas code and the Kloosterman sums is revealed. Our next task is to generalize $M(q)^\perp$ to a certain code, in which the hyper-Kloosterman sums are involved.

3. Hyper-Kloosterman code

In this section we define our new code and deduce some basic properties of it.

DEFINITION A.6 For any integer $m \geq 2$, we define the *hyper-Kloosterman code $C_m(q)$ of degree $m-1$* by the image of the map

$$\varphi_m : \mathbf{F}_q{}^m \to \mathbf{F}_p{}^{(q-1)^{m-1}}$$

given by

$$\varphi_m(\mathbf{a}) = \{\operatorname{Tr}(\mathbf{a}, \mathbf{x})\}_{\mathbf{x} \in (\mathbf{F}_q^\times)^{m-1}} . \qquad (A.16)$$

The symbol $\{\ \ \}_{\mathbf{x} \in (\mathbf{F}_q^\times)^{m-1}}$ represents a vector obtained by letting $\mathbf{x}$ run through the set $(\mathbf{F}_q^\times)^{m-1}$ (such a notation is often used in the literature on the trace codes). It is clear that $C_2(q) = M(q)^\perp$. And here are some basic properties of $C_m(q)$:

THEOREM A.4 (i) The code $C_m(q)$ is a linear $[(q-1)^{m-1}, mr]$ code unless $(m, q) = (2, 4)$ ($C_2(4)$ is a $[3, 2]$ code).
(ii) The Hamming weight of $\varphi_m(\mathbf{a})$ is given by

$$w(\varphi_m(\mathbf{a})) = \frac{1}{p}\left\{ (p-1)(q-1)^{m-1} - \sum_{s \in \mathbf{F}_p^\times} K_m(sa; q) \right\}. \qquad (A.17)$$

Proof. The linearity and the code length is trivial, and (ii) can be verified in a similar way to Theorem A.3. We give a sketch of proof for the dimension in (i).

First we list the vectors in $(\mathbf{F}_q^\times)^{m-1}$ as follows:

$$\boldsymbol{f}_0 = (1, 1, \cdots, 1)$$
$$\boldsymbol{f}_1 = (\alpha, 1, \cdots, 1)$$
$$\boldsymbol{f}_2 = (\alpha^2, 1, \cdots, 1)$$
$$\cdots\cdots$$
$$\boldsymbol{f}_{q-2} = (\alpha^{q-2}, 1, \cdots, 1)$$
$$\boldsymbol{f}_{q-1} = (1, \alpha, 1, \cdots, 1)$$
$$\boldsymbol{f}_q = (\alpha, \alpha, 1, \cdots, 1)$$

$$\cdots\cdots$$
$$\boldsymbol{f}_{(q-1)^2} = (1, \alpha^2, 1, \cdots, 1)$$
$$\cdots\cdots$$
$$\boldsymbol{f}_{(q-1)^{m-2}} = (1, \cdots, 1, \alpha)$$
$$\cdots\cdots$$
$$\boldsymbol{f}_{(q-1)^{m-1}-1} = (\alpha^{q-2}, \alpha^{q-2}, \cdots, \alpha^{q-2}).$$

Moreover we define $I(\boldsymbol{x})$ for $\boldsymbol{x} = (x_1, x_2, \cdots, x_{m-1}) \in (\mathbf{F}_q^\times)^{m-1}$ by

$$I(\boldsymbol{x}) = (x_1 x_2 \cdots x_{m-1})^{-1}.$$

Then we form the matrix $G_{m,q}$ as follows:

$$G_{m,q} = \begin{pmatrix} \boldsymbol{f}_0^{\mathrm{T}} & \boldsymbol{f}_1^{\mathrm{T}} & \cdots & \boldsymbol{f}_{(q-1)^{m-1}-1}^{\mathrm{T}} \\ I(\boldsymbol{f}_0) & I(\boldsymbol{f}_1) & \cdots & I(\boldsymbol{f}_{(q-1)^{m-1}-1}) \end{pmatrix}, \qquad (\mathrm{A}.18)$$

where $\boldsymbol{x}^{\mathrm{T}}$ is the transposed vector of $\boldsymbol{x}$. Now we consider the code $\bar{C}_m(q)$ over $\mathbf{F}_q$ with the generator matrix $G_{m,q}$. It is easy to see $C_m(q) = \mathrm{tr}\, \bar{C}_m(q)$. When $(m, q) = (2, 4)$, direct calculation shows $\dim C_2(4) = 2$. Otherwise, we can see that $\mathrm{rank}\, G_{m,q} = m$, and can verify by the definition of $G_{m,q}$, that the linear mapping $\mathrm{tr} : \bar{C}_m(q) \to \mathrm{tr}\,\bar{C}_m(q) = C_m(q)$ becomes injective (otherwise tr would be a constant mapping). The code $\bar{C}_m(q)$ has $q^m = p^{mr}$ vectors, and so does $C_m(q)$. ∎

EXAMPLE A.1 The generator matrix of $\bar{C}_3(4)$.

Let α be a primitive element of $\mathbf{F}_4$. Then $\alpha^3 = 1$ and $\alpha^2 + \alpha + 1 = 0$. Then we have

$$G_{3,4} = \begin{pmatrix} 1 & \alpha & \alpha^2 & 1 & \alpha & \alpha^2 & 1 & \alpha & \alpha^2 \\ 1 & 1 & 1 & \alpha & \alpha & \alpha & \alpha^2 & \alpha^2 & \alpha^2 \\ 1 & \alpha^{-1} & \alpha^{-2} & \alpha^{-1} & \alpha^{-2} & 1 & \alpha^{-2} & 1 & \alpha^{-1} \end{pmatrix}$$

$$= \begin{pmatrix} 1 & \alpha & \alpha^2 & 1 & \alpha & \alpha^2 & 1 & \alpha & \alpha^2 \\ 1 & 1 & 1 & \alpha & \alpha & \alpha & \alpha^2 & \alpha^2 & \alpha^2 \\ 1 & \alpha^2 & \alpha & \alpha^2 & \alpha & 1 & \alpha & 1 & \alpha^2 \end{pmatrix}. \qquad (A.19)$$

For the later use, we deduce more detailed expressions of the Hamming weights than (A.17). First suppose $a \in (\mathbf{F}_q^\times)^m$ and put $a = a_1 a_2 \cdots a_m$. Then change of variables immediately gives $K_m(a; q) = K_m(a; q)$, where

$$K_m(a; q) = \sum_{\substack{x_i \in \mathbf{F}_q^\times \\ x_1 x_2 \cdots x_m = a}} e(\mathrm{tr}\,(x_1 + x_2 + \cdots + x_m)), \qquad (A.20)$$

and it is also easy to see that

$$K_m(ba; q) = K_m(b^m a; q) \qquad (A.21)$$

for all $b \in \mathbf{F}_p^\times$ and $a \in (\mathbf{F}_q^\times)^m$.

On the other hand, if some entries of a are zero, then we know the explicit value of $K_m(a; q)$:

LEMMA A.5 Let $a = (a_1, a_2, \cdots, a_m) \in \mathbf{F}_q^m$ and suppose that t entries of a are zero $(1 \leq t \leq m)$. Then,

$$K_m(a; q) = (-1)^{m-t}(q-1)^{t-1}.$$

Proof. We can write

$$K_m(a; q) = \sum_{\substack{x_i \in \mathbf{F}_q^\times \\ x_1 x_2 \cdots x_m = 1}} e(\mathrm{tr}\,(a_1 x_1 + a_2 x_2 + \cdots + a_m x_m)). \qquad (A.22)$$

Since the right hand side is symmetric with respect to $a_1, a_2, \cdots a_m$, we can assume without loss of generality that

$$a_1, \cdots, a_{m-t} \in \mathbf{F}_q^\times \quad \text{and} \quad a_{m-t+1} = \cdots = a_m = 0.$$

Then we have

$$K_m(a; q) = \sum_{x_1, x_2, \cdots, x_{m-1} \in \mathbf{F}_q^\times} e(\mathrm{tr}\,(a_1 x_1 + a_2 x_2 + \cdots + a_{m-t} x_{m-t}))$$

$$= \sum_{x_{m-t+1} \in \mathbf{F}_q^\times} \cdots \sum_{x_{m-1} \in \mathbf{F}_q^\times}$$

$$\left\{ \sum_{x_1 \in \mathbf{F}_q^\times} e(\mathrm{tr}\,(a_1 x_1)) \cdots \sum_{x_{m-t} \in \mathbf{F}_q^\times} e(\mathrm{tr}\,(a_{m-t} x_{m-t})) \right\}$$

$$= \sum_{x_{m-t+1} \in \mathbf{F}_q^\times} \cdots \sum_{x_{m-1} \in \mathbf{F}_q^\times} \left\{ \sum_{x \in \mathbf{F}_q^\times} e(\mathrm{tr}\,(x)) \right\}^{m-t}$$

$$= (-1)^{m-t}(q-1)^{t-1}. \quad \blacksquare$$

If we combine (A.17), (A.21) and Lemma A.5, we obtain the following:

THEOREM A.6 Let $a = (a_1, \cdots, a_m) \in \mathbf{F}_q{}^m$, and we define two integers W_a and w_t by

$$W_a = \frac{1}{p}\left\{(p-1)(q-1)^{m-1} - \sum_{s \in \mathbf{F}_p^\times} K_m(s^m a; q)\right\} \quad (a \in \mathbf{F}_q^\times) \quad \text{(A.23)}$$

and

$$w_t = \frac{p-1}{p}\left\{(q-1)^{m-1} - (-1)^{m-t}(q-1)^{t-1}\right\} \quad (1 \le t \le m), \quad \text{(A.24)}$$

respectively. Then for any $\varphi_m(a) \in C_m(q)$, we have the following:

$$w(\varphi_m(a)) = \begin{cases} W_a, & \text{if } a \in (\mathbf{F}_q^\times)^m \text{ and } a := a_1 \cdots a_m, \\ w_t, & \text{if } t \text{ entries of } a \text{ are zero.} \end{cases} \quad \text{(A.25)}$$

This theorem shows that we know the explicit value of $w(\varphi_m(a))$ with at least one zero-entry in a, but not of $w(\varphi_m(a))$ with $a \in (\mathbf{F}_q^\times)^m$.

The following proposition is straightforward:

PROPOSITION A.7 If we denote by $A(w)$ the number of the codewords $\varphi_m(a)$ such that $w(\varphi_m(a)) = w$, we have

$$A(W_a) = (q-1)^{m-1} \quad \text{for all } a \in \mathbf{F}_q^\times, \quad \text{(A.26)}$$

$$A(w_t) = \binom{m}{m-t}(q-1)^{m-t} \quad (1 \le t \le m), \quad \text{(A.27)}$$

where $\begin{pmatrix} m \\ r \end{pmatrix}$ is the binomial coefficient.

REMARK A.3 The number of the codewords $\varphi_m(a)$ such that $a = a_1 a_2 \cdots a_m \neq 0$ is

$$\sum_{a \in \mathbf{F}_q^\times} A(W_a) = (q-1)^m. \qquad (A.28)$$

In the following sections, we deduce some properties of $C_m(q)$.

4.   Quasi-cyclic property

In the last section we introduced the hyper-Kloosterman code $C_m(q)$, which is a generalization of a cyclic code, the Kloosterman code. The Hamming weights of the codewords of $C_m(q)$ can be expressed by the hyper-Kloosterman sums $K_m(a; q)$, which are the generalization of $K(a, b; q)$, obtained by increasing the number of variables. Then what will happen to the code $C_m(q)$ by this generalization ? The answer is the quasi-cyclic property. So we begin this section by introducing the notion of a quasi-cyclic code:

DEFINITION A.7 A code C is called s- *quasi-cyclic* if

$$c_{n-s+1} \cdots c_n c_1 c_2 \cdots c_{n-s} \in C \qquad (A.29)$$

holds for every codeword $c_1 c_2 \cdots c_n$.

This property depends on the permutation of the coordinates, but if we take $G_{m,q}$ as the generator matrix, we can realize $C_m(q)$ as a quasi-cyclic code:

THEOREM A.8 The code $C_m(q)$ is $(q-1)^{m-2}$- quasi-cyclic.

Proof. Take a codeword $(a_1, a_2, \cdots, a_m)G_{m,q}$ of $\overline{C}_m(q)$ and apply a cyclic shift of $(q-1)^{m-2}$ digits to it. Then we can verify that the resulting vector is

$$(a_1, a_2, \cdots, a_{m-2}, a_{m-1}\alpha^{q-2}, a_m\alpha^{-(q-2)})G_{m,q}$$

and it is an element of $\overline{C}_m(q)$. Thus $\overline{C}_m(q)$ is $(q-1)^{m-2}$-quasi-cyclic, and so is $C_m(q)$. ∎

EXAMPLE A.2 $C_3(4)$.

We can obtain the following table of the codewords of $C_3(4)$:

Weight 0:	100000001	100001101	101000011	110100111	100101111
000000000		010100110	011101000	011111001	111011010
	Weight 4:	001010011	000011101	000111111	001011111
Weight 2:	011100100	100011100	110011000	111010110	111110100
000010001	110001001	001110001	000110011	001111101	111000111
000100010	101010010	010101010	011000110	010111011	011011011
000001100	011001010	100110010		111100101	110110110
010001000	101100001	010011001	Weight 6:	111001011	101101101
100010000	110010100	001101100	110111010	100111110	011110101
001100000	100100011	000101110	101001111	111111000	101011110
001000010	001001110	110000101	101111100	111101001	110101011
010000100	010010101	101110000	011010111	010110111	

Since this is the case $q = 4$ and $m = 3$, it is 3-quasi-cyclic. Take for example, the codeword 000010001 (the first one of weight 2). Move the last 3 digits of it to the beginning and shift the remainder to follow them. Then we get another codeword 001000010 (the seventh one of weight 2). The same procedure to 001000010 will produce 010001000, which is the fourth one. This also holds for all other codewords.

Quasi-cyclic codes have been investigated by many authors since Townsend-Weldon [34], but it seems that no one has ever considered the codes of our type, that is, quasi-cyclic trace codes. We will also be able to deduce quasi-cyclic generalization of other trace codes described by roots of polynomials.

5.   Weight distribution

We investigate the distribution of the Hamming weights of $C_m(q)$. One of the merits of considering the trace code is that we can know the distribution of the Hamming weights from that of the corresponding exponential sums. We apply this principle to our code. In this section we assume $p - 1 | m$. Then the Hamming weight of the type (A.23) can be expressed by a single hyper-Kloosterman sum, since $s^m = 1$ for all $s \in \mathbf{F}_p$:

$$w(\varphi_m(\boldsymbol{a})) = \frac{p-1}{p} \left\{ (q-1)^{m-1} - K_m(a; q) \right\}. \tag{A.30}$$

Moreover, the number of codewords with $\boldsymbol{a} \notin (\mathbf{F}_q^\times)^m$ is much smaller than that of those with $\boldsymbol{a} \in (\mathbf{F}_q^\times)^m$ (see Proposition A.7 and its remark). Hence under the condition $p-1|m$, the distribution of the hyper-Kloosterman sums immediately yields that of the Hamming weight. To know the distribution of the hyper-Kloosterman sums, we borrow a certain result from algebraic geometry, the Deligne-Katz theorem. We introduce some lemmas and notations to state it.

For the hyper-Kloosterman sums of the form (A.20), there is a well-known estimate called the Deligne bound:

THEOREM A.9 (DELIGNE)

$$|K_m(a; q)| \leq mq^{\frac{m-1}{2}}. \tag{A.31}$$

See [30]. Especially the case $m = 2$ is called the Hasse-Weil bound:

$$|K_2(a; q)| \leq 2\sqrt{q}. \tag{A.32}$$

The following formulation is due to Katz [13, Chapter 13]. From now on we suppose that m is even. We put $g = m/2$ and

$$T(m) := [-\pi, \pi)^g \quad \text{(the g-dimensional torus)}. \tag{A.33}$$

We define an equivalence relation in $T(m)$: for two elements $\Theta = (\theta_1, \cdots, \theta_g)$ and $\Psi = (\psi_1, \cdots, \psi_g)$ in $T(m)$, we write $\Theta \sim \Psi$ if and only if

$$(\theta_1, \cdots, \theta_g) = (\pm\psi_{\sigma(1)}, \cdots, \pm\psi_{\sigma(g)}) \tag{A.34}$$

for some $\sigma \in S_g$ (the symmetric group). In other words, we identify Θ and Ψ when they are mapped to each other by permutation of the entries or elementwise multiplication by -1. We consider the quotient set $T(m)/\sim$.

EXAMPLE A.3 (i) $m = 2$.
When $m = 2$, then $T(m) = [-\pi, \pi)$. And $x \sim -x$ for any $x \in [-\pi, \pi)$. Therefore $T(m)/\sim = [0, \pi]$ (note that $-\pi$ and π denote the same point in $T(m)$).

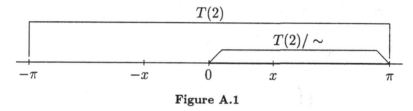

Figure A.1

(ii) $m = 4$.
When $m = 4$, then $T(m) = [-\pi, \pi)^2$. If we take a general point $(x, y) \in T(m)$, then there are 8 equivalent points in $T(m)$. Therefore $T(m)/\sim$ in this case becomes the inside and the edges of the triangle illustrated in the figure below:

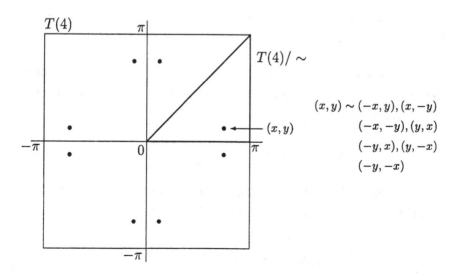

Figure A.2

With these settings, we can state the Deligne-Katz theorem on the distribution of the hyper-Kloosterman sums:

THEOREM A.10 (DELIGNE-KATZ) Let m be even. Then for all $a \in \mathbf{F}_q^\times$, there exists $\Theta(a) = (\theta_1(a), \cdots, \theta_g(a))$ in $T(m)/\sim$, such that

$$\frac{K_m(a,q)}{q^{(m-1)/2}} = 2(\cos \theta_1(a) + \cdots + \cos \theta_g(a)). \tag{A.35}$$

When $q \to \infty$, the points of $\{\Theta(a)\}_{a \in \mathbf{F}_q^\times}$ are uniformly distributed with respect to the measure

$$d\mu(m) = \frac{4^{g^2}}{2^g g!} \frac{1}{(2\pi)^g} \prod_{i=1}^{g} \sin^2 \theta_i$$

$$\cdot \prod_{1 \le i < j \le g} \sin^2 \left(\frac{\theta_i - \theta_j}{2}\right) \sin^2 \left(\frac{\theta_i + \theta_j}{2}\right) d\theta_1 \cdots d\theta_g. \tag{A.36}$$

More precisely for any test function f on $T(m)/\sim$, we have

$$\frac{1}{q-1} \sum_{a \in \mathbf{F}_q^\times} f(\Theta(a)) = \int_0^\pi f(\Theta) d\mu(m) + O\left(\frac{1}{\sqrt{q}}\right). \quad (q \to \infty) \tag{A.37}$$

Proof. This is a result from Katz [13, 13.5.3]. See also Deligne [5, 3.5.7]. ∎

Using this result, we can deduce the distribution theorem for the Hamming weights of $C_m(q)$. For a codeword $x = \varphi_m(a)$, we normalize the weight $w(x)$ by

$$w(x) \mapsto Z(x) = \frac{pw(x) - (p-1)(q-1)^{m-1}}{m(p-1)q^{(m-1)/2}}. \tag{A.38}$$

From (A.30) and the Deligne bound (Theorem A.9), we have $Z(x) \in [-1, 1]$ for $x = \varphi_m(a)$ such that $a \in (\mathbf{F}_q^\times)^m$. Indeed,

$$Z(x) = \begin{cases} \dfrac{-1}{mq^{(m-1)/2}} K_m(a; q), & \text{if } w(x) = W_a, \\[2ex] \dfrac{(1-q)^{t-1}}{mq^{(m-1)/2}}, & \text{if } w(x) = w_t. \end{cases} \tag{A.39}$$

Let f be a test function (i.e. continuous and compactly supported) on $\mathbf{R}$. Then (A.26), (A.27) and (A.39) give

$$\sum_{x \in C_m(q)} f(Z(x)) = \sum_{t=1}^{m} \binom{m}{m-t} (q-1)^{m-t} f\left(\frac{(1-q)^{t-1}}{mq^{(m-1)/2}}\right)$$

$$\text{(A.40)}$$

$$+ \sum_{a \in \mathbf{F}_q^{\times}} (q-1)^{m-1} f\left(\frac{-1}{mq^{(m-1)/2}} K_m(a; q)\right)$$

Remark. We are mainly interested in the interval $[-1, 1]$, so it seems sufficient to assume the continuity of f on $[-1, 1]$, but we assume a stricter condition for f because $Z(w_t)$ is not always contained in $[-1, 1]$.

To pass to the distribution of the normalized hyper-Kloosterman sums

$$-\frac{K_m(a; q)}{mq^{(m-1)/2}}$$

from that of $\Theta(a)$'s, we use the following lemma:

LEMMA A.11 Let X and Y be compact Hausdorff spaces, let μ a positive Borel measure on X with $\mu(X) = 1$, and $T : X \mapsto Y$ be a continuous mapping. If the sequence $(x_n)_{n \in \mathbf{N}}$ is uniformly distributed in X with respect to μ, then $(Tx_n)_{n \in \mathbf{N}}$ is uniformly distributed in Y with respect to $T^{-1}\mu$, where $T^{-1}\mu$ is defined by

$$T^{-1}\mu(E) = \mu(T^{-1}(E))$$

for all $E \subset Y$.

Proof. This is Exercise 1.10 in Chapter 3 of Kuipers-Niederreiter [16].

∎

Combining (A.40), Theorem A.10 and Lemma A.11, we get the main theorem on the distribution of the Hamming weights:

THEOREM A.12 Let m be even and $p - 1 | m$. When $q \to \infty$, the normalized Hamming weights $\{Z(x); \ x \in C_m(q)\}$ are uniformly distributed in $[-1, 1]$ with respect to a measure $d\nu_m$ induced by $\mu(m)$. Precisely, for any continuous and compactly supported function $f : \mathbf{R} \to \mathbf{C}$, we have

$$\frac{1}{q^m} \sum_{x \in C_m(q)} f(Z(x)) = \int_{-1}^{1} f(y) d\nu_m(y) + O\left(\frac{1}{\sqrt{q}}\right), \quad (q \to \infty) \quad (A.41)$$

where the measure $d\nu_m(y)$ is determined by

$$\int_{-1}^{1} f(y) d\nu_m(y) := \int_{T(m)/\sim} \tilde{f}(\Theta) d\mu(m), \quad (A.42)$$

$$\tilde{f}(\Theta) := f\left(\frac{2}{m}(\cos \theta_1 + \cdots + \cos \theta_g)\right). \quad (A.43)$$

EXAMPLE A.4 (i) $m = 2$.

This case was first pointed out by Lachaud [17] (in the binary case). The measure $d\mu(2)$ is the so-called Sato-Tate measure:

$$d\mu(2) = \frac{2}{\pi} \sin^2 \theta d\theta. \quad (A.44)$$

We have

$$\frac{K_2(a; q)}{2\sqrt{q}} = \cos \theta_a \quad (A.45)$$

and the induced measure is

$$d\nu_2(y) = \frac{2}{\pi} \sqrt{1 - y^2} dy. \quad (A.46)$$

Therefore (A.41), the precise statement of uniform distribution becomes

$$\frac{1}{q^2} \sum_{x \in C_2(q)} f(Z(x)) = \int_{-1}^{1} f(y) \cdot \frac{2}{\pi} \sqrt{1 - y^2} dy. \quad (A.47)$$

This formula is easily obtained from (A.37) by change of variables $\Theta = \arccos y$, because both $T(2)/\sim$ and $[-1, 1]$ are one-dimensional.

(ii) $m = 4$.

In this case we have

$$d\mu(4) = \{\sin \theta_1 \sin \theta_2 (\cos \theta_1 - \cos \theta_2)\}^2 d\theta_1 d\theta_2. \quad (A.48)$$

Therefore we determine $d\nu_4(y)$ by

$$\int_{-1}^{1} f(y)d\nu_4(y) := \int_{T(4)/\sim} f\left(\frac{1}{2}(\cos\theta_1 + \cos\theta_2)\right) d\mu(4). \qquad (A.49)$$

The distribution of the Hamming weights follows the measure $\nu_4(y)$ thus defined.

6.   Minimum distance of $C_m(q)$

We consider the minimum distance of $C_m(q)$. In this section we restrict ourselves to the case $p - 1 | m$. Under this condition, the Hamming weight of the codeword $\varphi_m(a) \in C_m(q)$ $(a = (a_1, a_2, \cdots, a_m) \in F_q^m)$ takes one of the following values:

$$w_s = \frac{1}{2}\{(q-1)^{m-1} + (-1)^{m-s+1}(q-1)^{s-1}\}, \quad (1 \le s \le m)$$

$$W_a = \frac{1}{2}\{(q-1)^{m-1} - K_m(a;q)\}, \quad (a \in F_q^\times),$$

where $s = \#\{i | 1 \le i \le m, a_i = 0\}$ and $a = a_1 a_2 \cdots a_m \in F_q^\times$. Clearly if $m \ge 3$, we have

$$\min_{\substack{1 \le s \le m \\ w_s \ne 0}} w_s = w_{m-2} = \frac{1}{2}\{(q-1)^{m-1} - (q-1)^{m-3}\}.$$

Therefore the biggest one of $(q-1)^{m-3}$ and $K_m(a;q)$ gives the minimum distance. We know the estimate of $K_m(a;q)$, the Deligne bound (Theorem A.9) . So basically, we compare two values $(q-1)^{m-3}$ and $mq^{(m-1)/2}$.

The goal of this section is the following:

THEOREM A.13 Suppose $p - 1 | m$ and let $d(m, q)$ be the minimum distance of $C_m(q)$. Then we have the following:
(i)

$$d(m, q) = \frac{p-1}{p}\{(q-1)^{m-1} - (q-1)^{m-3}\}$$

$$\text{if} \begin{cases} m \ge 6, & q = 8, q \ge 16, \\ m \ge 8, & q = 4, 9. \end{cases}$$

(ii)

$$d(m,q) \geq \tilde{d}(m,q) = \frac{p-1}{p} \left\{ (q-1)^{m-1} - m \cdot q^{\frac{(m-1)}{2}} \right\}$$

$$\text{if} \quad \begin{cases} 3 \leq m \leq 7, & q = 4, \\ 2 \leq m \leq 6, & q = 9, \\ 2 \leq m \leq 5, & q = 8, q \geq 16, \end{cases}$$

except for $(m,q) = (2,4)$.

REMARK A.4 We give the explicit value for the missing case $(m,q) = (2,4)$: $d(2,4) = 2$.

First we prove an easy lemma:

LEMMA A.14 Suppose $(q-1)^{m_0-3} \geq m_0 q^{\frac{m_0-1}{2}}$ for some m_0 and q $(m_0, q \in \mathbf{Z}, m_0 \geq 2, q \geq 4)$. Then for all $m \geq m_0$ we have $(q-1)^{m-3} \geq m q^{\frac{m-1}{2}}$.

Proof. It is easy to see

$$\max_{\substack{m \geq 2 \\ q \geq 4}} \frac{m+1}{m} \cdot \frac{\sqrt{q}}{q-1} \leq 1.$$

From this we have

$$\frac{m_0+1}{m_0} \cdot \frac{\sqrt{q}}{q-1} \cdot m_0 q^{\frac{m_0-1}{2}} \leq (q-1)^{m_0-3},$$

which gives the conclusion by induction on m. ∎

Proof of Theorem A.13. (i) First we consider the function

$$f(x) = (x-1)^3 - 6x^{\frac{5}{2}}.$$

We can verify $f(x) \geq 0$ if $x \geq 49$. This together with Lemma A.14 yields

$$(q-1)^{m-3} \geq m q^{\frac{m-1}{2}} \qquad \text{(A.50)}$$

for all prime powers $q \geq 49$ and all $m \geq 6$. There are 7 prime powers (not primes themselves) less than 49 ($q = 4, 8, 9, 16, 25, 27, 32$). For these, we can see by direct calculation that (A.50) holds when

$$m \geq 8, \quad q = 8, 9, 16, 25, 27, 32,$$
$$m \geq 13, \quad q = 4.$$

Moreover, computer calculation of $K_m(a; q)$ leads to the conclusion (see the tables at the end of this section).

(ii) When $m = 2$, we have $w_1 = \{(q-1)+1\}/2$ and $w_2 = 0$. Clearly $w_1 \geq \{(q-1) - 2\sqrt{q}\}/2$, and all the nonzero Hamming weights are greater than $\tilde{d}(2, q)$ for $q \neq 2$ (when $q = 4$, $\tilde{d}(2, q) < 0$ and estimation becomes trivial). When $3 \leq m \leq 5$, the smallest w_s but 0 is w_{m-2}, and we can easily verify $(q-1)^{m-3} \leq mq^{(m-1)/2}$ for $q \geq 4$. Thus we have proved (ii) for $2 \leq m \leq 5$, $q \geq 4$ $((m, q) \neq (2, 4))$. The remaining cases are due to computer calculation of $K_m(a; q)$. ∎

As this theorem shows, for large values of m, some of the vectors $a \in (\mathbf{F}_q)^m$ with zero entries give the codewords of the minimum distance, but for small m's, they are given by $a \in (\mathbf{F}_q^\times)^m$. In these cases the minimum distance is expressed by the sums $K_m(a; q)$. The exact value of $K_m(a; q)$ is hard to determine, so there is no other way than to evaluate $d(m, r)$ by the value $\tilde{d}(m, r)$.

Here are the results of numerical calculation. In the tables below, $\max K_m(a; q)$ means $\max_{a \in \mathbf{F}_q^\times} K_m(a; q)$.

$\underline{q = 4}$

m	6	7	8	9	10	11	12
$(q-1)^{m-3}$	27	81	243	729	2187	6561	19683
$mq^{(m-1)/2}$	192	448	1024	2304	5120	11264	24576
$\max K_m(a; q)$	43	85	171	341	683	1365	2731

$\underline{q = 8}$ $\underline{q = 9}$

m	6	7
$(q-1)^{m-3}$	343	2401
$mq^{(m-1)/2}$	1086.1...	3584
$\max K_m(a; q)$	247	713

m	6
$(q-1)^{m-3}$	512
$mq^{(m-1)/2}$	1458
$\max K_m(a; q)$	584

$q = 16$

m	6	7
$(q-1)^{m-3}$	3375	50625
$mq^{(m-1)/2}$	6144	28672
$\max K_m(a;q)$	1519	9745

$q = 27$

m	6
$(q-1)^{m-3}$	17576
$mq^{(m-1)/2}$	22727.9...
$\max K_m(a;q)$	6398

$q = 32$

m	6	7
$(q-1)^{m-3}$	29791	923521
$mq^{(m-1)/2}$	34755.7...	229376
$\max K_m(a;q)$	11359	60577

REMARK A.5 Are the hyper-Kloosterman codes good ? To see the efficiency of $C_m(2^r)$, let us use the Gilbert-Varshamov curve as a criterion:

$$y = 1 - H_2(x),$$

where $H_2(x) = -x \log_2 x - (1-x) \log_2(1-x)$. Suppose $C_m(2^r)$ is a $[n, k, d]$-code. Then we can prove

$$\frac{k}{n} \geq 1 - H_2\left(\frac{d}{n}\right)$$

if $2 \leq m \leq 5$ and r is sufficiently large. This shows that good binary hyper-Kloosterman codes would exist in the range $2 \leq m \leq 5$.

7.　A divisibility theorem for Hamming weights

Considering the divisibility of the Hamming weight is sometimes important. For example (in the binary case), a code C is called doubly-even if $4|w(x)$ for all $x \in C$. Doubly-even codes are very important if they are self-dual. Since the hyper-Kloosterman sums $K_m(a;q)$ are algebraic integers in the cyclotomic field $\mathbf{Q}(\zeta_p)$ ($\zeta_p = e^{2\pi i/p}$), we can deduce a divisibility theorem for the Hamming weights of $C_m(q)$ by employing some knowledge of algebraic number theory.

In this section we assume $p - 1|m$ and put

$$K = \mathbf{Q}(\zeta_p), \quad L = \mathbf{Q}(\zeta_p, \zeta_{q-1}). \tag{A.51}$$

We denote the rings of integers in K and L by O_K and O_L, respectively. As usual, for an algebraic number field M and $\eta \in M$, the norm and the trace of η are defined by

$$N_{M/\mathbf{Q}}\eta = \prod_{\sigma \in \mathrm{Gal}(M/\mathbf{Q})} \eta^\sigma,$$

$$Tr_{M/\mathbf{Q}}\eta = \sum_{\sigma \in \mathrm{Gal}(M/\mathbf{Q})} \eta^\sigma.$$

Then we have the following:

THEOREM A.15 Let $m \geq 2, r \geq 2, q = p^r$ and $p-1|m$. Then $K_m(a; q) \in$ **Z** and we have

$$(q - 1)K_m(a; q) \equiv (-1)^m \pmod{p^{\frac{m}{p-1}}}. \qquad (A.52)$$

Proof. When $p - 1|m$, (A.30) is true, and from this formula we know $K_m(a; q) \in \mathbf{Z}$. In fact, since $w(\varphi_m(\boldsymbol{a})) \in \mathbf{Z}$, we have $K_m(a; q) \in \mathbf{Q}$ from (A.30). On the other hand $K_m(a; q)$ is an algebraic integer in K. Thus we get $K_m(a; q) \in \mathbf{Z}$ because $K_m(a; q) \in O_K \cap \mathbf{Q}$.

To deduce the congruence, we first appeal to Fourier analysis over finite fields. Recall α is a primitive element of $\mathbf{F}_q$ and suppose $a = \alpha^j$ ($0 \leq j \leq q - 2$). Then we have the following formula:

$$K_m(a; q) = \frac{1}{q - 1} \sum_{k=0}^{q-2} \bar{\psi}_k(\alpha^j)G(\psi_k)^m, \qquad (A.53)$$

where

$$\psi_k(\alpha^j) = e^{2\pi i k j/(q-1)} \quad \left(\begin{array}{c}\text{the multiplicative}\\ \text{character of } \mathbf{F}_q\end{array}\right), \qquad (A.54)$$

$$G(\psi_k) = \sum_{t=0}^{q-2} \psi_k(\alpha^t)e(\mathrm{tr}\,(\alpha^t)) \text{ (the Gauss sum for } \psi_k), \quad (A.55)$$

and $\bar{\psi}_k$ is the complex conjugate of ψ_k. The formula (A.53) is known as the Fourier inversion of $K_m(a; q)$. See for example, inxKatz [13, Chapter 4]. Since $G(\psi_0) = -1$, we get from (A.53),

$$(q - 1)K_m(a; q) = (-1)^m + \sum_{k=1}^{q-2} \bar{\psi}_k(\alpha^j)G(\psi_k)^m. \qquad (A.56)$$

For $1 \leq k \leq q-2$, we consider the divisibility of $G(\psi_k)$ in O_L.

In the theory of cyclotomy, it is well known that $\pi := 1 - \zeta_p$ is a prime in O_K, $\pi | p$ and $N_{K/\mathbf{Q}}\pi = p$. Since $e(\operatorname{tr}(\alpha^t)) = \zeta_p^{\operatorname{tr}(\alpha^t)}$, we have

$$
\begin{aligned}
e(\operatorname{tr}(\alpha^t)) &= (1+\pi)^{\operatorname{tr}(\alpha^t)} \\
&= 1 + \operatorname{tr}(\alpha^t)\pi + \cdots \\
&= 1 + \xi_t \pi
\end{aligned}
\tag{A.57}
$$

for some $\xi_t \in O_L$. Therefore in O_L,

$$
\begin{aligned}
G(\psi_k) &= \sum_{t=0}^{q-2} \psi_k(\alpha^t)(1 + \xi_t \pi) \\
&= \sum_{t=0}^{q-2} \psi_k(\alpha^t) + \left(\sum_{t=0}^{q-2} \psi_k(\alpha^t)\xi_t\right)\pi \\
&\equiv 0 \ (\operatorname{mod} \pi),
\end{aligned}
\tag{A.58}
$$

since $\sum_{t=0}^{q-2} \psi_k(\alpha^t) = 0$ when $k \neq 0$. Thus we have from (A.56)

$$
(q-1)K_m(a;q) - (-1)^m \equiv 0 \ (\operatorname{mod} \pi^m)
\tag{A.59}
$$

in O_L. Note that the left hand side of (A.59) is in $\mathbf{Z}$ and $\pi \in O_K$. Considering these facts, if we take the norm $N_{L/\mathbf{Q}}$, we have

$$
\{(q-1)K_m(a;q) - (-1)^m\}^{[L:\mathbf{Q}]} \equiv 0 \ (\operatorname{mod} p^{[L:K]m})
\tag{A.60}
$$

in $\mathbf{Z}$. We get the desired congruence by this formula since $[L : \mathbf{Q}] = [L : K](p-1)$. ∎

REMARK A.6 Recall the sum in (A.23) and note that

$$
\sum_{s \in \mathbf{F}_p^{\times}} K_m(s^m a; q) = Tr_{K/\mathbf{Q}} K_m(a;q).
\tag{A.61}
$$

We can show $K_m(a;q) \in \mathbf{Z}$ when $p-1 | m$ from this identity. Indeed we have $K_m(s^m a; q) = K_m(a;q)$ under this condition and hence

$$
Tr_{K/\mathbf{Q}} K_m(a;q) = (p-1)K_m(a;q),
\tag{A.62}
$$

which is equivalent to $K_m(a;q) \in \mathbf{Q}$.

We get a divisibility theorem for the Hamming weights in synthesis of Theorem A.15, formulas (A.24) and (A.30):

THEOREM A.16 When $(p-1)|m$, we have

$$\min(p^{r-1}, p^{\frac{m}{p-1}-1})|w(x)$$

for all $x \in C_m(q)$.

Proof. For the codeword $\varphi_m(a)$ such that $a = a_1 a_2 \cdots a_m \neq 0$, we have from (A.30) and Theorem A.15,

$$
\begin{aligned}
pw(\varphi_m(a)) &= pw_a = (p-1)\left\{(q-1)^{m-1} - K_m(a;q)\right\} \\
&\equiv (p-1)\left\{(q-1)^{m-1} - qK_m(a;q) - (-1)^{m-1}\right\} \\
&\qquad\qquad\qquad\qquad (\bmod\ p^{\frac{m}{p-1}}) \\
&\equiv (p-1)\left\{p^{r(m-1)} - (m-1)p^{r(m-2)} + \cdots \right. \\
&\qquad \left. +(m-1)p^r(-1)^{m-2} - p^r K_m(a;q)\right\} \quad (\bmod\ p^{\frac{m}{p-1}}) \\
&\equiv 0 \quad (\bmod\ \min(p^r, p^{\frac{m}{p-1}})).
\end{aligned}
$$

We have the same congruence for the codeword $\varphi_m(a)$ such that $a_1 a_2 \cdots a_m = 0$ by direct calculation, using (A.24). Thus we obtain the theorem. ∎

Especially in the binary case,

COROLLARY A.17 For any $m \geq 2$ and $r \geq 2$, we have

$$\min(2^{r-1}, 2^{m-1})|w(x) \qquad\qquad (\text{A.63})$$

for all $x \in C_m(2^r)$.

Therefore, if $m \geq 3$ and $r \geq 3$, then $C_m(2^r)$ is doubly-even. But if $m = 2$ or $r = 2$, there really exist $C_m(2^r)$'s which are not doubly-even, as the following example shows:

EXAMPLE A.5 Weight distribution for some $C_m(2^r)$'s.

_r = 2_

$C_3(2^2)$ (length=9)

Weight	0	2	4	6	Total
Number	1	9	27	27	64

$C_4(2^2)$ (length=27)

Weight	0	8	12	14	16	18	Total
Number	1	27	54	108	54	12	256

$C_5(2^2)$ (length=81)

Weight	0	30	36	40	42	46	54	Total
Number	1	81	90	405	270	162	15	1024

_r = 3_

$C_2(2^3)$ (length=7)

Weight	0	2	4	6	Total
Number	1	21	35	7	64

$C_3(2^3)$ (length49)

Weight	0	16	24	28	Total
Number	1	49	294	168	512

$C_4(2^3)$ (length343)

Weight	0	160	168	172	184	196	Total
Number	1	1029	637	1372	1029	28	4096

_r = 4_

$C_2(2^4)$ (length=15)

Weight	0	4	6	8	10	Total
Number	1	30	60	105	60	256

$C_3(2^4)$ (length=225)

Weight	0	96	108	112	116	120	Total
Number	1	450	900	675	900	1170	4096

References

[1] E. Bannai; Codes over finite rings and finite abelian groups (a survey). In: Proc. Conf. on Algebraic Geometry, Number Theory, Coding Theory and Cryptography. Univ. of Tokyo, Jan.1998, 98-107.

[2] K. Chinen and T. Hiramatsu; Hyper-Kloosterman sums and their applications to the coding theory, Appl. Algebra Engrg. Comm. Comput., 12 (2001), 381-390.

[3] H. Cohen; Trace des opérateur de Hecke sur $\Gamma_0(N)$, in Séminaire de Théorie des Nombres, exp. no. 4, Bordeaux, 1976-1977.

[4] J. H. Conway and N. J. A. Sloane; Sphere Packings, Lattices and Groups, Springer, 1988.

[5] P. Deligne; La Conjecture de Weil II, Pub. Math. I. H. E. S. 52 (1980), 137-252.

[6] M. Deuring; Die Typen der Multiplikatorenringe elliptischer Funktionenkörper, Abh. Math. Semi. Hansischen Univ., 14 (1941), 197-272.

[7] W. Ebeling; Lattices and Codes, Vieweg, 1994.

[8] E. Grosswald; Representations of Integers as sums of Squares, Springer, 1985.

[9] S. H. Hansen; Rational Points on Curves over Finite Fields, Lecture Notes Series, No.64, Aarhus Univ., 1995.

[10] R. Hartshorne; Algebraic Geometry, Graduate Texts in Math. 52, Springer, 1977.

[11] T. Hiramatsu; Uniform distribution of the weights of the Kloosterman codes, SUT J. Math., 31 (1995), 29-32.

[12] T. Høholdt, J. H. van Lint and R. Pellikaan; Algebraic geometry codes, Chapter 10 in Handbook of Coding Theory, Edited by V. S. Pless and W. C. Huffman, Elsevier Science B. V., 1998.

[13] N. M. Katz; Gauss Sums, Kloosterman Sums, and Monodromy Groups, Ann. Math. Studies 116, Princeton Univ. Press, 1988.

[14] H. D. Kloosterman; On the representation of numbers in the form $ax^2 + by^2 + cz^2 + dt^2$, Acta Math., 49 (1926), 407-464.

[15] _____; Asymtotische Formeln für die Fourierkoeffizienten ganzer Modulformen, Abh. Math. Semi. Hamburg Univ., 5 (1927), 337-352.

[16] L. Kuipers and H. Niederreiter; Uniform Distribution of Sequences, John Wiley & Sons, 1974.

[17] G. Lachaud; Distribution of the weights of the dual of the Melas code, Discrete Math., 79 (1989/90), 103-106.

[18] J. H. van Lint; Introduction to Coding Theory, Springer, 1982.

[19] J. H. van Lint and G. van der Geer; Introduction to Coding Theory and Algebraic Geometry, Birkhäuser, DMV Seminar, Band 12, 1988.

[20] J. B. Little; Applications to coding theory, in Proc. of Symposia in Applied Math., 53 (1998), 143-167.

[21] F. J. MacWilliams and N. J. A. Sloane; The Theory of Error-Correcting Codes, North-Holland, 1977.

[22] C. M. Melas; A cyclic code for double error correction, IBM J. Res. Devel., 4 (1960), 364-366.

[23] A. J. Menezes; Elliptic Curve Public Key Cryptosystems, Kluwer, 1993.

[24] C. Moreno; Algebraic Curves over Finite Fields, Cambridge Univ. Press, 1991.

[25] R. Pellikaan, B.-Z. Shen and G. J. van Wei; Which linear codes are algebraic-geometric ?, IEEE Trans. IT, 37 (1991), 583-602.

[26] H. Petersson; Über die Entwicklungskoeffizienten der automorphen Formen, Acta Math., 58 (1932), 169-215.

[27] O. Pretzel; Error-Correcting Codes and Finite Fields, Clarendon Press, Oxford, 1992.

[28] R. Schoof and M. van der Vlugt; Hecke Operators and the weight distributions of certain codes, J. Combin. Theory, Ser.A, 57 (1991), 163-186.

[29] R. Schoof; Families of curves and weight distributions of codes, Bulletin AMS, 32 (1995), 171-183.

[30] Séminair de géometrie algébrique du Bois-marie, SGA $4\frac{1}{2}$, Lecture Notes in Math., Vol. 569, Springer, 1977.

[31] J. P. Serre; A Course in Arithmetic, Springer, 1985.

[32] S. A. Stepanov; Codes on Algebraic Curves, Kluwer Academic / Plenum Publishers, 1999.

[33] H. Stichtenoth; Algebraic Function Fields and Codes, Springer, 1993.

[34] R. L. Townsend and E. L. Weldon, Jr.; Self-orthogonal quasi-cyclic codes, IEEE Trans. IT, 13 (1967), 183-195.

[35] M. A. Tsfasman, S. G. Vladut and Th. Zink; Modular curves, Shimura curves and Goppa codes, better than Varshamov-Gilbert bound, Math. Nachr. 109 (1982), 21-28.

[36] J. L. Walker; Codes and Curves, Student Math. Library, Vol. 7, AMS and Institute for Advanced Study, 2000.

[37] J. Wolfmann; The weights of the dual code of the Melas code over $GF(3)$, Discrete Math., 74 (1989), 327-329.

[38] _____; The number of solutions of certain diagonal equations over finite fields, J. Number Theory, 42 (1992), 247-257.

[39] _____; New result of diagonal equations over finite fields from cyclic codes, Contemporary Math., 168 (1995), 387-395.

Index